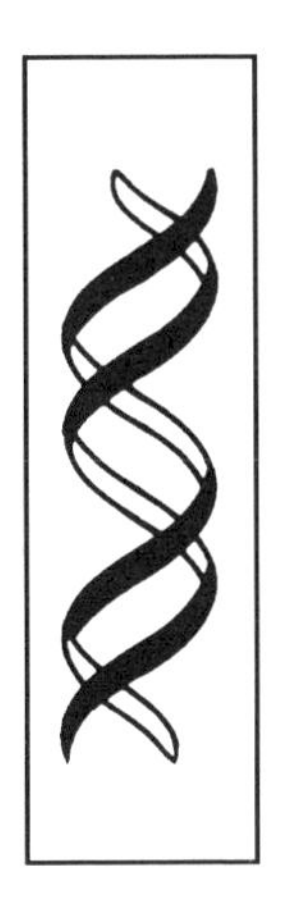

Gene Doping in Sports: The Science and Ethics of Genetically Modified Athletes

Advances in Genetics, Volume 51

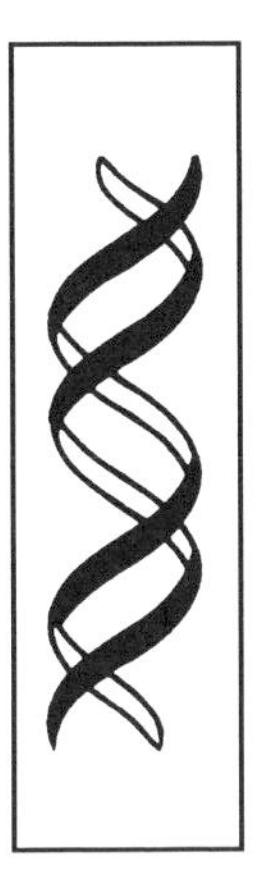

Gene Doping in Sports: The Science and Ethics of Genetically Modified Athletes

Angela J. Schneider
Faculty of Health Sciences
Health Sciences Addition UWO
The University of Western Ontario
London, Ontario, Canada

Theodore Friedmann
Center for Molecular Genetics
University of California at San Diego School of Medicine
La Jolla, California

AMSTERDAM • BOSTON • HEIDELBERG • LONDON
NEW YORK • OXFORD • PARIS • SAN DIEGO
SAN FRANCISCO • SINGAPORE • SYDNEY • TOKYO
Academic Press is an imprint of Elsevier

Elsevier Academic Press
525 B Street, Suite 1900, San Diego, California 92101-4495, USA
84 Theobald's Road, London WC1X 8RR, UK

This book is printed on acid-free paper. ♾

For all information on all Elsevier Academic Press publications
visit our Web site at www.books.elsevier.com

ISBN-13: 978-0-12-017651-9
ISBN-10: 0-12-017651-3

PRINTED IN THE UNITED STATES OF AMERICA
06 07 08 09 9 8 7 6 5 4 3 2 1

Contents

Foreword

The science of genetic enhancement and its application across a broad range of social phenomena require the most careful consideration and the development of criteria and protocols that will provide a framework within which experimentation, testing and implementation can proceed. The use of these techniques has an impact on how we define humanity, thereby necessitating close consideration of the ethics.

This work focuses on ethical issues as they relate to sport, including the regulations for prevention of doping generally and genetic enhancement in particular.

The pace of scientific development in gene therapy in three decades has been remarkable, and the ongoing growth of knowledge is far more likely to be logarithmic than arithmetic. Apart from the science itself, some of the medical questions are little short of revolutionary. Imagine the idea of not merely treating the symptoms of a disease, but instead, using genes to attack and eliminate the root of the imperfection or disease itself. There have been setbacks, which will always occur in experimental medicine as knowledge expands and the risks emerge or are identified.

What is essential is to be aware of the need for an ethical framework, responsible oversight and transparent protocols. There must also be a distinction between legitimate therapy and other possible applications of the new scientific technology. It is critical to define the distinction between therapy and enhancement, from treating the debilitation of muscular dystrophy to simply enhancing muscles, from triggering muscle growth to taking off the brakes provided by myostatin, where the goal of what may appear to be treatment is enhancement. Those who maintain that glucocortico steroids should be prohibited will note a potential argument in support of that position in chapter 4.

Perhaps the most perceptive observation regarding the standards for research and experimentation in the genetic domain is whether they would conform to the standard requirements for human research if they were carried out openly. The Nuremberg and Helsinki codes, for example, provide useful and thoughtful guidelines for ethical research. Where a subject could not be in a position to give a consent within such a structured framework, it would follow that the subject could not give an informed consent and any consequent research would be unethical.

One bright element in the whole picture of genetic enhancement in sport is that the leading scientists and the sports movement have taken pre-emptive steps to work together as the science develops, for purposes of collaborating to establish the necessary definitional aspects of the problem and to develop tests to determine whether gene transfer technology has been applied. This situation is quite different from the one that was allowed to develop with respect to the use of performance-enhancing drugs and methods, starting in the late 1950s, where the sport authorities were blind (sometimes willfully blind) to the existence of the problem and have been playing an expensive game of catch-up ever since. The initiative of the International Olympic Committee to create the World Anti-Doping Agency (WADA) and establish for it a governance structure with equal representation of governments and athletic interests, combined with a secure budget (similarly equally funded), has been an important milestone in the fight against doping in sport. The definition of doping, contained in the World Anti-Doping Code adopted by WADA and applied throughout the Olympic sports and some professional sports and adopted by the UNESCO International Convention Against Doping in Sport, has been expanded to include genetic manipulation.

No one can afford to ignore the fact that gene therapy is a reality. There will be ongoing efforts to wrestle with the practical applications and the larger questions of what is to be, or should be, treated, but its time has come. Those concerned with determining whether it may have been improperly applied, particularly in sport, will have to be ready to operate on the assumption that there may well be clandestine applications, deliberately outside the international norms and guidelines, and for the express purpose, not of therapy, but of gaining an advantage in competition through enhancement.

The authors of this useful and timely study are two preeminent thinkers–one a leading scientist engaged at the forefront of an exciting branch of science, and the other, a former Olympic medalist whose academic focus has been on the plethora of ethical issues arising in sport. They present a snapshot of a scientific and ethical continuum, identifying the key principles that govern how the development should proceed in a brief and informative manner.

Richard W. Pound

Preface

This book has arisen out of growing concerns in the scientific and athletic communities that the tools of genetic manipulation that have been developed for the treatment of human genetic diseases are going to be applied to normal human athletes to enhance sports performance. This presumption is part of the larger scientific concern regarding genetically based enhancement of many human traits that do not determine human illness or deformation, but rather that govern so much about what it means to be human—our physical traits, personalities, and emotions. All of these traits are determined to a greater or lesser extent by both our genes and environment, just as are most human diseases—cancer, cardiovascular disease, neuropsychiatric disease, diabetes, and virtually all other human afflictions. And to the extent that we are learning very quickly how to manipulate the genes responsible for disease and develop definitive new treatments; we are beginning to appreciate that those same tools and methods are applicable to many other human traits and that we are likely to be able to modify many of our other physical, cognitive, and personality traits in addition to correcting our diseases.

Among those traits now becoming vulnerable to directed change are those that determine how good athletes we can become—how strong we are, how fast we can run, how well we recover from injury, and how well we tolerate discomfort and pain. The scientific world has become aware that there is a broad area of genetic "enhancement" that must be defined and that the societal and ethical quandaries associated with such technologies must be defined and resolved. Similarly, the world of athletics—the athletes, their handlers, and their professional associations—are also aware of these developing scientific advances, and have come to envision ways in which they represent severe challenges to sports.

In response to that growing concern, the International Olympic Committee (IOC) convened a small working group of scientists, athletes, and ethicists in Lausanne to begin to examine the issues of genetics in sport doping. In 2002, the World Anti-Doping Agency (WADA) brought together a larger group of representatives in the scientific and athletic communities for several days to identify in greater detail the nature and severity of the problem for sports. The meeting was held in one of the world's premiere biomedical research centers—the Banbury Center of the Cold Spring Harbor Laboratory on Long Island, New York—and was the first such effort to allow the scientists, athletes, ethicists, and public policy leaders to develop a common language and understanding of the issues and begin to develop programs to counter them. One of the major objectives of the meeting was to organize informational and educational programs to bring the question of genetic enhancement and sport doping to the attention of more people in science, sports, and public policy. This book is a direct outgrowth of that goal. Our goal in this effort has been to make the fields of gene transfer, gene therapy, human experimentation and clinical research, genetic enhancement, and sports pressures understandable to a general audience of athletes, scientists, policy experts, and interested general readers.

The views that we present are our views and should not be considered to represent official positions of WADA. Nevertheless, we hope that it does present an accurate picture of the spirit of the Banbury Conference and of the subsequent concepts that have arisen from that initial meeting. During the preparation of this book, we have had continuing and enthusiastic cooperation and help from WADA. In particular, we wish to thank Richard Pound, Chairman of WADA, Prof. Arne Ljungqvist, member of the IOC, Chairman of the IOC Medical Commission and Chairman of the WADA Health, Medicine, and Research Committee and Dr. Olivier Rabin, Science Director of WADA.

Angela Schneider
Theodore Friedmann

1

The Problem of Doping in Sports

It is a safe bet that virtually all of the traits of living creatures are determined and shaped by a combination of genetic and environmental influences. This is just as true for human beings as it is for bacteria, trees, and frogs. Physical traits, such as our hair color and height, as well as our personalities, intelligence, talents, and emotional properties, are the result of the potential provided by our genes and the nurturing and shaping of that potential by the environment in which we develop and grow. Furthermore, the presumed mix of the effects of "nature" and "nurture"—genetics and our environment—applies not only to what makes us normal but also to the "abnormal" traits that cause so many human diseases and suffering. In many cases, such as sickle-cell anemia and cystic fibrosis, the genetic factors are predominant. These diseases result from defects—"mutations"—in single genes, and chances are high that people who have inherited abnormal genes will develop the clinical disease, no matter how the environment is manipulated. "Nature" is enough to cause the disease. For other diseases, such as, various forms of diabetes, most heart diseases and cancer, psychiatric diseases, and so on, there may be a number of genetic factors that have to interact with each other and that also are expressed in many different ways depending

Advances in Genetics, Vol. 51
0065-2660/06 $35.00
DOI: 10.1016/S0065-2660(06)51001-6

on the environmental influences. "Nature" interacts with "nurture" to bring about the clinical disease. In such cases, making changes in "nurture" may have a very significant effect on how that disease trait is expressed—hence, the reasonably effective treatments for many diseases with drugs, surgery, vaccines, and so on.

It is not only disease characteristics that are affected by the influence of genes and environment but athletic performance that is affected by many of these, mixed "nature"–"nurture" factors. Of course, to be successful in sports, one must have the physical talents and capabilities—the appropriate body build, keen eyesight, and rapid hand–eye coordination to satisfy the physical demands of the sports. But it is just as important for the underlying athletic talent and physical capability to emerge in an environment that fosters participation and training—an encouraging family with sufficient resources to support purchase of equipment and the expenses of trainers and a school environment that provides athletic opportunities. It is hard to imagine becoming a successful athlete without having both the physical characteristics as well as the milieu for those traits to develop.

The parts of this formula for success in sports can be, and often are, manipulated to enhance the likelihood for successful athletic performance. Only by ensuring the best possible training facilities and opportunities for athletes and providing the best medical support and care for them can we ensure that a sport maintains its important role in one's society in fostering a healthy sense of striving for and achieving lofty goals, building a rewarding sense of physical accomplishment in the young, and providing our society with the inspiration and entertainment of sports achievement. Enlightened programs exist in many parts of the world to encourage participation in sports, provide sports facilities in schools and communities for recreational participation in sports, as well as for training selected athletes in sports academies and other sports facilities for more professional and elite levels of sports. Such schools and centers provide equipment and space for training and competition and make available the expertise and time of trainers, nutritionists, and medical personnel to ensure the best care and nurturing of athletes. The goals of such facilities should be to provide broad access to all who would like to participate, while, at the same time, identifying and fostering the training of those who aim at higher levels of sports competition, possibly even at elite international and Olympic levels. Thus, our society openly and honestly manipulates some aspects of the environment of sports to improve performance.

Unfortunately, our society can manipulate other aspects of the sports environment in sly, duplicitous, surreptitious, dangerous, and venal ways, as indicated by the pervasive problem of drug abuse—doping—in sports. This is not a new phenomenon in sports, since it is recognized that athletes for centuries, as far back as ancient Greece and probably even earlier, have pursued victory by using any available means, including specialized ointments, lotions and salves, diets, and so on, openly or secretly, to achieve athletic success. In the past, many of these kinds of tools and tricks were more or less ineffective and of little use, other than for psychological support of the athlete or intimidation of competitors. However, the problems associated with doping in sports today are different from the mere psychological effects in the past because the potency of drugs in common use today give them the potential not only for having powerful and real effects on athletic performance effect but also for equally powerful but harmful physiological effects. Today's drugs are not yesterday's placebos. The use of potent and sometimes highly hazardous drugs by athletes to enhance their performance is common at all levels of sports training and competition, even among very young school children who aspire to excel in sports and therefore emulate their sports idols.

Our societies have come largely to accept the idea that this kind of manipulation of sports and sports performance is undesirable and requires control, regulation, and even banning. The considerations of devising bans on drug use in sports have been based on the following considerations.

1. Doping in sports is cheating and unfair. It can be argued that doping should be banned because it is cheating or unfair. The problem with this position is that usually an activity is considered "cheating" or "unfair" only when there is a rule prohibiting it. Without the rules, there may be no issue of cheating and unfairness. Others have suggested that an action can be inherently unfair even if there is no

rule prohibiting it, but a sport is inherently a rule-driven enterprise and without rules, questions of inherent unfairness and cheating would be very difficult to resolve in sports.

2. Doping harms the athletes. Here the suggestion is that doping should be banned to protect the athletes who dope. It may be argued that this rationale is paternalistic, inconsistent, and incomplete. It is paternalistic because we do not generally permit intrusion into the lives of competent adults under the guise of protecting them from perceived harms. It is inconsistent because a sport, in particular an elite-level sport, is often a hazardous enterprise. It is not clear why athletes should be protected from the harm that doping may inflict when we do not protect them from possible dangers of the training and performance of these same sports. The argument is incomplete because evidence for harm in doping is mixed. For example, while steroids in high doses may cause adverse side effects, steroids in relatively low doses probably do not. Autologous blood doping has not been shown to have any adverse side effects.
3. Doping harms even nondoping athletes. The argument based on the interests of other "clean" (nondoped) athletes is that doping can be coercive. Because doping may improve results, there is coercive pressure "to keep up" placed on those who wish to compete without doping. This argument has some merits, but is incomplete because an elite-level sport is already highly coercive. For example, when full-time training, attitude training, or diet control are shown to produce better results, everyone is forced to adopt these measures to keep up. It is unclear why doping is any more coercive and sufficiently so to warrant being banned than say, training 6–8 hours a day. On the other hand, the coercion argument has merit if it can be shown that doping detracts from what is important to sports. If sports, sporting excellence, and sporting contests are about testing skills, then it can be argued that the improved performance that comes with doping is not about that test of skill. If doping does not enhance the development of sports and sporting excellence, we can choose to reject it as being unnecessarily coercive, as compared to, for example, extended training that may improve one's skill at the contest at hand.

4. Doping harms society. This position is based on the assumption that doping harms others in society, especially children who see athletes as role models. This argument works in two ways. One, if children see athletes having no respect for the rules of the games they play, there will be a tendency to undermine respect for rules, and law, in general. The second version sees doping and athletic drug use as part of a wider social problem of drug use. The argument here is that if children see athletes using doping to attain sporting success, then other drugs may also be seen as a viable means to other ends. The limitations of this argument are that there are many things that we consider appropriate for adults but not for children. Alcohol and cigarettes are obvious examples, as is sex, and although we attempt to control these behaviors, we do not generally ban these substances or activities for adults because of their potential harmful effect on children.
5. Bans musts be enforceable and therefore require complex infrastructures for fair and just implementation. Many layers of intrusive and expensive testing and monitoring must be established to detect instances of forbidden drug use both in-competition and out–of-competition, to evaluate and judge the results of investigations, to provide mechanisms for appeal and to levy and enforce punishment—fines, suspensions and even at times criminal punishment. It is also necessary to develop and implement programs for responding to proven instances of doping—suspensions from competition, fines, etc. All these are expensive and potentially intrusive into the private lives of the athletes.
6. Doping represents a perversion of sports. It converts the beauty of sports, the glory of striving and achieving and outdoing physical limitations into "mere" biotechnology—into rewards for the most clever biological and genetic engineer instead of the swiftest, the strongest. Sport is a social construct that relies on an agreement on the part of all to abide by sets of rules, no matter how arbitrary, that are the agreed underlying principles. Taking part in organized sport is, contrary to some views, not a human right but rather a voluntary act by athletes who, by the simple act of agreeing to participate, agree to the rules and restrictions of that sport. Of course,

one of the human rights that athletes must be allowed to retain is the right to refuse to take part in doping control—no one, athlete or not, should be forced to accede to intrusive monitoring and screening programs without permission and without giving fully informed and voluntary consent. But if an athlete wishes to refuse to accede to doping control measures, he or she relinquishes the privilege of taking part in the sports competition.

7. Doping is unnatural and dehumanizing. The unnaturalness argument does not get very far for two reasons. First, we do not have a good account of what would count as "unnatural." Second, we are inconsistent. For any account of "natural" and "unnatural," few things designated "unnatural" would be permitted (e.g., spiked shoes) while other "natural" ones (e.g., testosterone) are on the banned list. The dehumanization argument is interesting but incomplete, since we do not have an agreed-upon conception of what it is to be human. Without this it is difficult to see why some practices should count as dehumanizing. We also have a problem with consistency. Some practices, such as "psychodoping," the mental manipulation of athletes using the techniques of operant conditioning, are not banned, whereas the reinjectionof one's own blood is. There is an attempt later to provide some framework for defining human excellence, thereby allowing us to see how thepursuit of athletic excellence can, and should, be limited in ways that exclude doping from the pursuit of sports.

I. THE DEVELOPMENT OF CURRENT PROGRAMS TO COUNTER DRUG ABUSE IN SPORTS

The international sports community has recognized for many years the dangers of all forms of doping, but it has been only in the recent past that serious and increasingly effective regulatory mechanisms have been put into place for the detection and control of drug-based doping in sports. It has been clear that, given the opportunity, athletes and their trainers and handlers will resort to many illicit techniques and substances to provide a competitive advantage in sports. One should only remember the pervasive and officially sanctioned and operated doping programs established in East Germany between 1970s and 1980s; how effective these were in the short term, and how harmful these were to the athletes in the long term. It seems very likely that the world of sports will continue to seek out new drugs and stealthy drug delivery methods and even gene-based enhancement to ensure victory in competition. Athletics represents one of the provinces of human activity, most susceptible to the application of existing and future advances in the field of human gene therapy, for the enhancement of nondisease human traits. Modern athletics is as much an entertainment as it is a sport and is sodden with huge amounts of money to assure the victories and records that the public demands. Athletic events are also some of the most powerful instruments for international politics. The prestige, nationalism, and jingoism compel our political institutions to demand victory. Finally, athletes are by nature risk-takers who are driven to compete, excel, and win, even at the cost of injury and other harm to themselves. But even worse, athletes are highly vulnerable to potentially harmful manipulation by dishonest and venal rogue trainers, sports technicians, and sports associations and federations who disregard the ideals of sports and the welfare of the athletes in the interests of victory at all cost. The fact that a sport is already filled with many pervasive drug-based forms of doping should convince even the most skeptical that all current and future advances in pharmacology, sports physiology, and sports medicine, whether based on ever more sophisticated drugs, gene transfer technology, or other still unrecognized technologies, will be applied to the world of sports and will almost certainly occur before the underlying technology is known to be effective and truly safe.

To improve the fight against this new potential kind of abuse, the International Olympic Committee (IOC) and national sports federations collaborated in 1998 to establish the World Anti-Doping Agency (WADA), an agency jointly funded by the IOC and cooperating nations and committed to develop programs for detection and control of athletic doping. It carries out its tasks by compiling and constantly updating a list of substances and methods that are inconsistent with the ideals of sports and that should be banned from athletic competition. It is also responsible for developing and validating new, scientifically sound detection assays and implementing effective international programs for in-competition and out-of-competition screening of athletes. In addition to this international effort, a number of countries, including the United States, have established national anti-doping agencies, similar to WADA, with the task of monitoring and controlling sports-doping issues at the national level and instituting research programs to develop even more effective assays for forbidden substances and methods. In the US agency, this national anti-doping effort is coordinated by the US Anti-Doping Agency. The WADA has implemented its program on drug control in sports by issuing and continually updating the world Anti-Doping Code, including a list of banned substances and methods, the latest of which is presented as an appendix to this volume.

II. A POTENTIAL NEW CHALLENGE FOR SPORTS: GENETIC DOPING

Although human beings have long had an intuitive knowledge of the mix of "nature" and "nurture" to treat or prevent disease and adapt ourselves physically, intellectually, and emotionally, we have relied entirely on manipulating the forces outside of ourselves. We have developed the tools of medicine—drugs, surgical methods, even psychotherapy—to heal the damage of disease. We think it is clear that we are on the verge of applying the same methods to change other human traits that have nothing to do with disease but rather affect other genetically based human characteristics—many of our physical attributes and at least some of our vital mental functions. Physical traits, such as muscle size and strength, blood circulatory properties, the efficiency of our energy utilization, and others, seem to be the most common targets of genetic manipulation just as these have been vulnerable to manipulation by "traditional" drug-based doping. If these methods are applied to sports, we propose that we will have to undergo a severe reevaluation of what we consider to be the essence and the value of sports.

The sports world has recognized the emergence of this potential problem by incorporating the concept of gene-based doping into the activities of some of the major national and international sports organizations, including the International Olympic committee (IOC), doping screening and monitoring agencies such as the WADA, the US Anti-Doping agency (USADA), the Netherlands Centre for Doping Affairs (NeCeDo), and others. The initial conference on the subject was convened by the IOC in Lausanne in June 2001 and that was followed in quick succession by international meetings and workshops organized by WADA, USADA, and NeCeDo. At the same time, the scientific world has also become aware of the issue and has presented a number of round-tables, symposia and lectures on the potential for gene-based doping in sports at scientific meetings. For instance, The American Association for the Advancement of Science (AAAS), the American Society for Gene Therapy (ASGT) have held major discussion of the subject at their major annual meetings, beginning in 2003. Finally, the subject has come to the attention of major ethics think tanks and governmental and policy bodies, as reflected most strongly by the examination of sport doping as one of the likely initial wedges in human genetic enhancement by the United States President's Council on Bioethics in 2003 and studies of gene transfer in sport doping by the Hastings Center, the major ethics think tank in the United States.

This book examines the scientific basis for this potential problem highlighting the scientific, ethical, and public policy considerations that the sports community, and society-at-large should understand and sort out. We aim in this book to present the background for potential gene-based doping and explain the ways in which this threat has emerged from the new area of medicine called "gene therapy." We will examine whether the tools of medically justifiable genetic modification of human subjects are appropriate for modification of human nondisease traits for what has been called "genetic enhancement," and to indicate our conclusion that a sport represents an area in which this development in human biology is likely to be first tested. We conclude that, for scientific, social, and ethical reasons, the use of gene transfer methods to enhance sports performance is to be condemned.

2

The Scientific Basis for Gene Therapy: A New Concept in Medicine

The explosion of genetic knowledge in the latter part of the twentieth century has given rise to new ways of thinking about human biology in health and disease. These advances have begun to deliver on their promise of new and more effective approaches to the prevention and treatment of human disease. This genetic revolution started long ago in the middle of the nineteenth century with the work of Gregor Mendel and the concepts of inheritance that emanated from his studies of the properties of peas and other plants in the monastery in Brno in the Austro-Hungarian Empire. Mendel discovered that the physical properties of these living systems are inherited in predictable and reproducible ways and that there are biological factors in the plants that carry these traits from one generation to the next.

Mendel's work was forgotten by the scientific community until the beginning of the twentieth century when his studies were "rediscovered"

Advances in Genetics, Vol. 51

0065-2660/06 $35.00
DOI: 10.1016/S0065-2660(06)51002-8

independently by several scientists and, for the first time, applied to human disease by the English clinician Sir Archibald Garrod during the first decade of the twentieth century. In his clinical work, Garrod was treating patients suffering from a number of familial conditions. The rules that Mendel had worked out in plants were found to be equally important to human beings in determining human disease. Garrod was struggling to understand and treat conditions, such as albinism (the absence of normal skin and hair pigmentation), pentosuria (a benign condition characterized by excessive excretion of the sugar pentose in the urine), alkaptonuria (a severe form of arthritis), and others. Garrod understood more by intuition than from formal demonstration that these conditions resulted from defects in the chemical factors that determined the mechanisms responsible for the production or metabolism of chemical molecules in the body. He coined the term "inborn errors of metabolism" to describe the underlying processes that led to the appearance of these and other inherited human conditions (Fig. 2.1).

Garrod's fortuitously chosen phrase "inborn errors of metabolism" is filled with prescience and insight. Every term in it represents an epochal advance in understanding of human biology. The word "inborn" reflected Garrod's understanding that it was precisely the human equivalent of genetic determinants, discovered by Mendel half a century earlier that underlay human disease as much as they determine whether a pea is wrinkled or smooth. "Errors" reveals Garrod's realization that the Mendelian factors, whatever they were, exist in an incorrect or altered form in inherited diseases and that they therefore produce aberrant physical properties in humans (e.g., disease and "metabolism") showed that Garrod understood that the action of these factors, normal or abnormal, was to drive metabolism in the human body. Here, through the work of Garrod, as described in his famous Croonian Lectures on Inborn Errors of Metabolism delivered to the Royal College of Physicians of London in 1908, emerged the modern chemical understanding of human disease—chemical pathology.

The structure or composition of the factors discovered by Mendel, and first applied to human disease, was not known for many years. At the time of conceptualization of "inborn errors," neither Garrod nor any other scientist had identified the chemical nature of the normal

Figure 2.1. Sir Archibald Garrod, who laid the groundwork for the medical discipline known as molecular pathology and who originated the phrase "inborn errors of metabolism." His clinical work was the first to connect the newly rediscovered concepts of Mendelian genetics with human disease.

or aberrant forms of the factors that determined heredity or produced genetic disease. Immense progress was made in the mechanisms of heredity in a number of laboratories, most notably in the "fly room" at Columbia University in New York, where Thomas Hunt Morgan and his colleagues were studying breeding in what came to be one of the most important model systems in genetics—the fruit fly. During the first decades of the twentieth century, Morgan and his students and colleagues moved Mendel's theoretical concept of inheritance factors to a physical reality by showing that changes at specific locations (mutations) along the chromosome produced heritable changes in the fly.

But even these and many other advances did not identify what the genes were—what substances and molecules they were composed of and what their structure was. Many scientists thought that the responsible molecule might be protein; some thought that genes might consist of nucleic acids that were known to be present in the chromosome, whereas others assumed all sorts of incorrect and fanciful chemical compositions and structures for the gene. The discovery of the true chemical nature of the gene was made in 1944 through the work of Oswald Avery and his group at the Rockefeller Institute in New York. Avery and his team found that they could permanently and heritably change the infectivity properties of a strain of the bacterium pneumococcus by introducing deoxyribonucleic acid (DNA) from a different strain. More precise confirmation came from subsequent studies by other scientists, but the Avery's experiments left little doubt that genes were composed of one particular form of nucleic acid, DNA, and that it was the information in this molecule that determined the properties of living things and their succeeding generations.

With the discovery of the precise three-dimensional structure of DNA by James Watson and Francis Crick in Cambridge in 1953 and the elucidation of the genetic code by several groups of molecular biologists came the first real functional understanding of DNA—how it replicated itself and encoded the genetic information that is passed from one generation to another. The "golden age" of molecular biology from 1953 through the 1970s and through the subsequent decades of one startling advance after another, have now culminated in the total sequence characterization of the human genome. Through all of these advances we know that: (a) the genetic information of living systems, from microbes to humans, is contained in the sequence of four simple basic purine and pyrimidine chemical-building blocks; (b) many naturally occurring differences in that sequence account for normal variations among individuals of a species; and (c) some differences in the sequence of genes, naturally occurring or induced by environmental influences, cause severe disruption of normal metabolic, cellular functions to cause disease. The work of Garrod illuminated the simplest of these kinds of genetic diseases—those resulting from errors or mutations in single genes. These kinds of diseases include cystic fibrosis, sickle-cell anemia, and many others, including the disorders studied by Garrod. However, now we know that the most common human diseases and those responsible for the greatest health burdens in our society (e.g., cancer, cardiovascular disease, diabetes, and most neurological and neuropsychiatric disorders, and so on) are the result of multiple genes interacting with each other.

I. HOW HUMAN GENETIC DISEASE OCCURS AND HOW IT IS TREATED

We know from the epochal work of Garrod and the century of advances that followed, many human diseases are caused by errors in our genetic material. Although we have acquired knowledge about genetic basis of many human diseases, the sophistication of that knowledge has not been matched by the sophistication of treatment of gene-based diseases. From the beginning of medical history to the present, therapies have almost always been targeted toward the effects of the aberrant genes rather than the genes themselves. There are, to be sure, several exceptions to that generalization. Surgery is a form of therapy that targets a structural abnormality and therefore can be thought of definitive therapy aimed at cause rather than effects of a disease. Similarly, antibiotic and antiviral therapies for infectious diseases, target the underlying causative agents rather than merely the metabolic derangements caused by infection. Similarly, tissue and cell transplantation can be imagined to be aimed at the causes of disease and the elegance and efficacy of bone marrow transplantation to treat and even cure life-threatening diseases, such as leukemias, some forms of thalassemia, and so on, which are additional examples of treatments aimed at the cause rather than the result of disease processes. In the cases of the best understood genetic diseases, the disorders that Garrod termed "inborn errors" and that are caused by defects in single genes, treatment is indirect and aimed to correct the metabolic consequences of the underlying genetic defect rather than the causative factor—the genetic defect itself.

The general approach to the treatment of disease is presented in Fig. 2.2 in the context of our modern understanding of the genetic components of most human diseases. The underlying defect is indicated

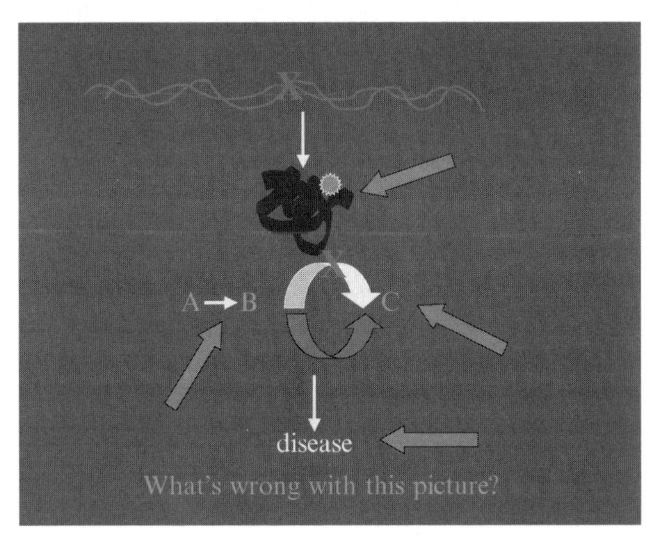

Figure 2.2. A schematic representation of the nature of genetic disease and the approaches generally taken for treatment. The two Xs indicates the site of the causative genetic error in the DNA and the resulting defective metabolic step. A, B, and C represent parts of the aberrant metabolic pathway responsible for disease and the arrows indicate the targets for most current forms of therapy. The clear message from this scheme is that treatment of human disease at the level of the underlying defect, the gene, is likely to be more effective and more definitive than treatment aimed at the metabolic consequences of the genetic defect.

by the X chromosome in the two-stranded representation of the DNA that comprises the total genetic information, the "genome," of a living being, as well as by the metabolic pathway that depends on the product of that gene. The function of the genes can be imagined to serve as a blueprint for the production of proteins—the functional molecules that carry out the cellular and metabolic functions specified by the genes. If the genes are normal, the proteins coded by them are normal and functional in their role of driving all of the processes of a living cell. If the gene has an error or "mutation," the encoded protein may not function properly and therefore the cellular functions that depend on that protein may not be performed appropriately, leading to cellular abnormalities and resulting disease.

Figure 2.3. The schematic approach to the damage caused by genetic disease. Treatments aimed at the consequences of the defects rather than the defects are almost always inadequate and destined to fail because they fight the losing ballet of cleaning up after the damage is done and while the damage is ongoing.

If one imagines genetic disease to be analogous to the damage caused by floods downstream from a broken dam, therapy has traditionally emphasized cleaning up the flood rather than fixing the dam. The dam continues to leak and produces further mess, requiring ever more frenetic, and ultimately futile, cleanup work in the flooded and damaged areas (Fig. 2.3). Flood cleanup cannot substitute for fixing the dam. Similarly, while symptomatic treatment certainly helps in many disorders, it would be far better to correct the responsible genetic defect.

Treatment for most human diseases has traditionally been aimed at the defective gene products, by replacing the products that are produced by the abnormal cellular and metabolic steps, by replacing damaged organs and tissues, and so on. For instance, the metabolic product insulin is given to replenish the insulin that fails to be produced adequately in diabetes. Human growth hormone is provided by injection to children whose growth is stunted because growth hormone production in their own pituitary glands is inadequate. Cancers are treated by drugs that interfere with many normal metabolic processes that regulate how quickly and how often cells divide and how quickly they grow. In these cases, the treatments are aimed not at the cause, but at the effect of the underlying defect. In other words, traditional disease treatments are symptomatic and are aimed at every step of the disease process rather than at the underlying genetic defect (Figs. 2.2 and 2.3).

What is wrong with this picture is that the treatments are aimed everywhere except at the most logical target—the genetic error. The theory that therapies are usually aimed at the wrong targets may help to explain why treatment of many diseases is not very effective, even with the most modern drugs and techniques in medicine today.

In the early 1970s, the concept began to emerge that to treat or prevent disease by altering the function of the responsible underlying gene would be more effective rather than aiming at the "downstream" effects of the genetic defect, as illustrated in Fig. 2.4. This change in thinking was due partly to the explosion that was occurring in understanding the structure of DNA, the nature of genes as segments of DNA that encode protein products, the nature of mutations and their role in human disease, and the development of methods to isolate specific genes, to characterize them, and to "splice" molecules of DNA together to create entirely new combinations of genetic information.

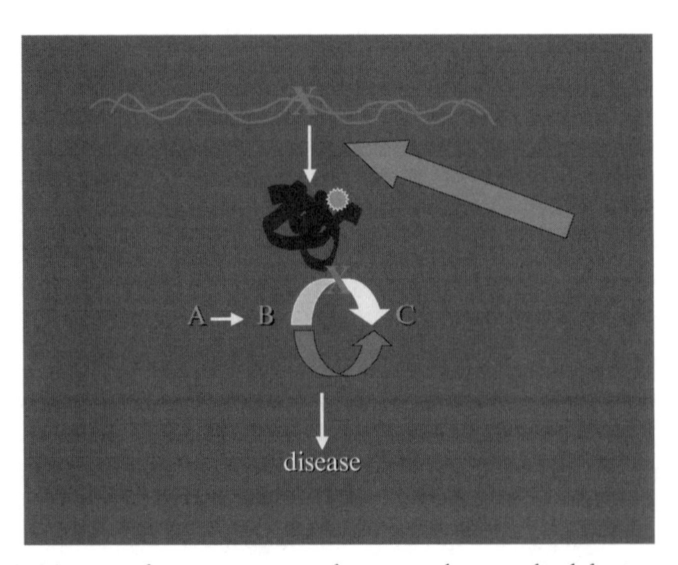

Figure 2.4. More specific treatment target for genetic disease—the defective gene itself.

II. GENE TRANSFER: A NEW APPROACH TO THERAPY

In the late 1960s and early 1970s, during the explosion of knowledge of molecular genetics, techniques emerged that began to promise precisely the new kind of approach to therapy. The genetic components of human disease were recognized, detailed mechanisms of gene expression were obtained, animal models of more and more human diseases were available for study in the laboratory, and early techniques for transferring foreign genes into human and other mammalian cells appeared. Inevitably, the connection of these advances in basic genetics with the need for definitive new methods of disease therapy was made, and the concept of "gene therapy" was born. The genetic material was not merely a static and fixed instruction book for human biology and human traits, but rather a potentially therapeutic material that could be used either as a drug or a biological agent to correct disease. This concept of gene therapy for human disease was first put forward clearly in 1972 in a publication by Theodore Friedmann and Richard Roblin in *Science* magazine.

From this conception more than three decades ago, it was clear that there were many difficult scientific, ethical, and policy obstacles to overcome before such techniques could be applied to humans for treatment of disease. Because of several ill-conceived and inappropriate applications of highly immature gene transfer technologies to human patients, an extensive set of oversight and regulation mechanisms have been put into place over the years to ensure that work in this extremely difficult area of clinical research will be carried out with scientific care and rigor and with maximal care for the welfare of patients and experimental subjects. Institutions that receive federal funding area required to have an Institutional Review Board (IRB) and an Institutional Biosafety Committee (IBC) to evaluate the scientific rationale and the safety of all studies dealing with human subjects. Further oversight is provided by the Office of Biotechnology Activities at the National Institutes of Health through its Recombinant DNA Advisory Committee (RAC) that reviews every study dealing with gene transfer in human subjects in the United States. Finally, the Food and Drug Administration (FDA) wields true regulatory authority for all human studies and is the final authority

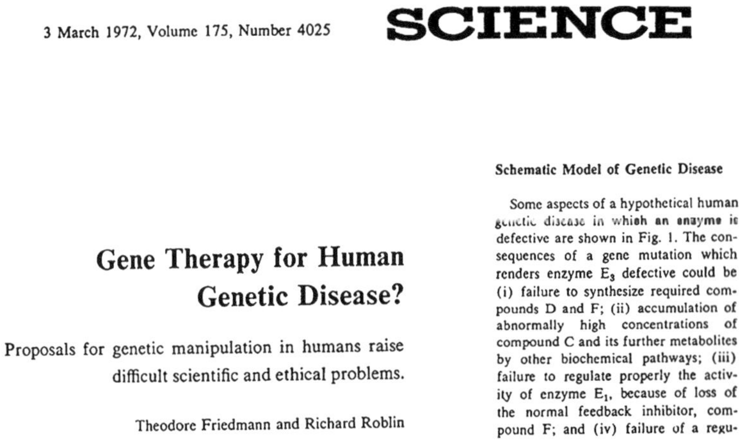

3 March 1972, Volume 175, Number 4025 **SCIENCE**

Gene Therapy for Human Genetic Disease?

Proposals for genetic manipulation in humans raise difficult scientific and ethical problems.

Theodore Friedmann and Richard Roblin

Schematic Model of Genetic Disease

Some aspects of a hypothetical human genetic disease in which an enzyme is defective are shown in Fig. 1. The consequences of a gene mutation which renders enzyme E_3 defective could be (i) failure to synthesize required compounds D and F; (ii) accumulation of abnormally high concentrations of compound C and its further metabolites by other biochemical pathways; (iii) failure to regulate properly the activity of enzyme E_1, because of loss of the normal feedback inhibitor, compound F; and (iv) failure of a regu-

Figure 2.5. The first page of the article in which the need for human gene therapy and its underlying concepts were first explicitly presented. The article also laid out the potential uses of virus vectors for gene transfer and also described many of the still relevant issues of regulation and oversight, premature human clinical application, and the potential for use in nondisease settings to enhance normal human traits.

on whether a gene transfer study is allowed to proceed in the United States. The combined work of these agencies and oversight bodies over the past 15 years or so has been to examine more than 700 proposed human gene transfer studies, involving thousands of research subjects and patients, and many crucial lessons have been learned about the efficacy and safety of the current techniques (see later chapters) Fig. 2.5.

Genes generally require some kind of vehicle or "vector" in order to be transferred into human cells and tissues with maximal efficiency. The DNA itself, in its naked form, can get into human cells to some extent. Even more efficient methods of gene transfer involve wrapping the DNA into a virus particle that has been disabled through genetic engineering methods. This removes the potentially harmful and dangerous viral genes and replaces them with a potentially therapeutic gene. Over the past decades, such incapacitated viral "vectors" have been made from many kinds of starting viruses, including some relatively innocuous viruses, as well as some potentially hazardous viruses, such as

retrovirus 1981-2 random
integration, insertional mutagenesis?

adenovirus

adeno-associated virus

liposomes

naked DNA

Figure 2.6. A variety of some of the most widely used and popular viral and nonviral "vectors" for gene transfer. The retrovirus vectors were the first truly effective gene transfer vectors, but some of their properties, such as their permanent integration, into the genes of the cells and their inability to "infect" some kinds of cells, such as brain neurons, led to the development of gene transfer vectors from other parent viruses, namely adenoviruses and adeno-associated viruses. Liposomes are not viruses but rather are artificially prepared bags of fatty molecules that are capable of transferring gene into human cells.

pathogenic adenoviruses, cancer-causing mouse leukemia viruses, and the human immunodeficiency virus (HIV). In all cases, the parent viruses are no longer able to produce their usual diseases because their genes are no longer present in the modified gene transfer vector. Some of these kinds of viral and nonviral vectors are demonstrated in Fig. 2.6.

Gene transfer vectors have been constructed from many other viruses (e.g., herpes viruses, pox viruses, flu viruses, even HIV—the cause of acquired immune deficiency syndrome [AIDS]). In such cases, the genes of the viruses themselves are removed or otherwise inactivated, leaving the virus with the only function of acting as a moving van, or in more surreptitious terms, a Trojan horse, to sneak foreign genes into the cells in a way that will be completely harmless to the cell.

III. CLINICAL DELIVERY: HOW THERAPEUTIC GENE TRANSFER VECTORS CAN BE INTRODUCED INTO HUMAN BEINGS

Two principal methods have been developed to introduce gene transfer vectors into research subjects and patients. These methods are now called the "*ex vivo*" and "*in vivo*" methods and differ by whether the gene transfer vector is introduced directly into the tissues of the patient or research subject ("*in vivo*") or alternatively into cells of the patients in the laboratory followed by the reintroduction of the genetically corrected cells into the patient ("*ex vivo*") (Fig. 2.7). For instance, cells can be obtained from the bone marrow of a patient suffering from a disease that prevents normal production of blood cells. These cells can then be modified genetically by exposure to a gene transfer vector in the laboratory and the corrected cells can then be transfused back into the patient. Under these conditions, the corrected cells make their way into the defective bone marrow again, set up residence there, and replace the abnormal cells. In another *ex vivo* approach, skin cells can be obtained from a patient through a very small skin biopsy, grown in culture dishes in the laboratory, genetically modified as discussed earlier, and then put back into the patient in a number of tissues, including the brain where they can produce a product to correct a genetic deficiency in that organ (Fig. 2.6).

The use of a patient's own cells for this *ex vivo* approach to gene transfer has the great advantage of avoiding the immune rejection of cells derived from other donors, whether they be closely related donors, unrelated donors, or possibly even nonhuman cells. In the latter cases, cells could be first put into microcapsules that prevent contact of the "foreign" cells with the host immune system that would cause an immune rejection response by the host and elimination of the therapeutic cells.

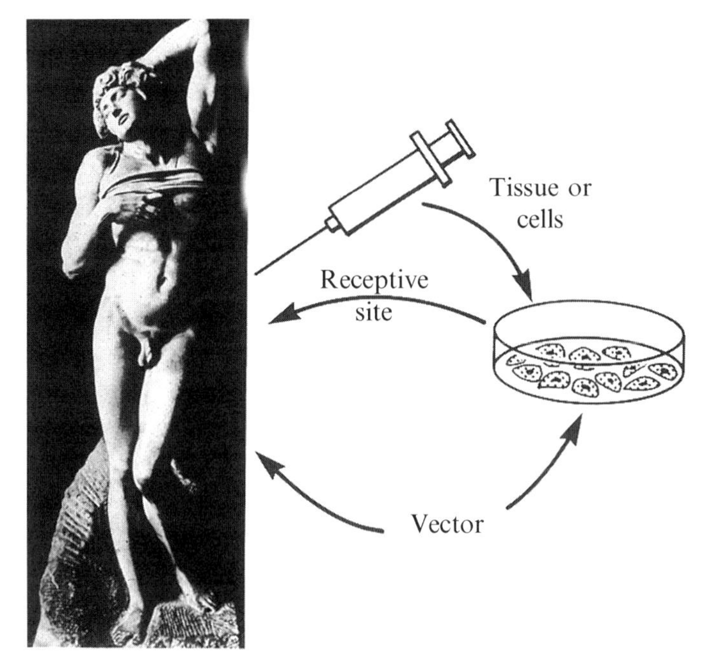

Figure 2.7. The *ex vivo* and *in vivo* approaches to vector and gene transfer into human beings. The vector can be applied to cells that have been derived from a patient and manipulated in cell culture conditions in the laboratory or directly to tissues and organs in the patient. Both methods are being pursued in many clinical gene transfer studies in human.

In the case of the *in vivo* approach to gene transfer, a vector or even a "naked" piece of DNA can be introduced by injection directly into a tissue to transfer a foreign gene into the cells of that tissue. One of the early surprises of the *in vivo* approach of gene transfer was that injection of DNA directly into skeletal muscle was a relatively effective way of causing a genetic modification of muscle cells and of providing prolonged expression of a foreign gene to alter muscle function. Most of these early studies involved genes that are needed specifically in the muscle cells themselves and that are involved in diseases such as muscular dystrophy. But muscle is also an excellent platform from which one can deliver many kinds of gene products into the blood stream for delivery to the entire body, and many foreign genes other than those producing muscle-specific functions have also been introduced into and expressed by muscle, including genes encoding growth factors, erythropoietin, and others. To achieve more efficient levels of gene transfer than those possible with naked DNA alone, virus vectors have also been injected directly into many tissues and organs *in vivo*, including many kinds of cancer, skeletal and even heart muscles, the brain, joints, the skin, the liver, the lungs and other parts of the airway, and many other sites.

IV. FINE-TUNING THE EXPRESSION OF THE FOREIGN GENE: NOT TOO MUCH OR TOO LITTLE, NOT AT THE WRONG TIME

Depending on the tissue and the kind of vector used, the level and duration of new gene expression can vary greatly from very short duration, or low-level expression, to very prolonged and sometimes robust expression, and the choice of the technique depends on the therapeutic needs. Is it necessary to have a constant supply of the new gene product, sudden and transient burst of very high-level expression, a high sustained level of expression, and a minimal but prolonged degree of expression? Should the expression of the foreign gene be regulated so that it is "turned on" at some times and "off" at other times? These are crucial issues for normal gene expression and cellular function because genes in the human body are not merely "on" or "off," but rather are subject to exquisite and ever-changing regulation to satisfy changing physiological needs. Too much or too little gene expression, or expression at the wrong times, can be very harmful. Normal expression of foreign genes that would duplicate the expression of the body's own genes would ideally require the incorporation into the vectors of small segments of DNA that respond to these body signals for increased or decreased gene expression, but the technology for this kind of regulation is very early in its development and is just the beginning to be extended from interesting laboratory methods to techniques in living creatures *in vivo*.

V. REVERSING OR STOPPING THE EFFECTS OF FOREIGN GENES IF THINGS GO WRONG

What can be done if things go seriously wrong with a gene transfer procedure, such as the activation of a cancer gene and the escape of the genetically altered cells? We know now that our tools are still too immature to prevent serious harm to some people taking part in gene transfer experiments. It seems clear that the more effective we become with today's tools and methods, the greater are the chances of serious and harmful side effects. Therefore, it is prudent to incorporate some procedures to remove the foreign gene or even to kill the cells carrying the transferred gene. In some cases, it might be possible to remove the cells surgically, especially if the gene transfer is performed into skin, an accessible muscle, or other tissue that can be accessed readily and can be removed with impunity. In those cases in which the foreign gene is placed into the body in the form of a small capsule from which the therapeutic gene product can be secreted into a local tissue or into the blood stream, the capsule can be removed.

One such method of selective cell removal involves the incorporation into the gene transfer vector of a "suicide" gene—a second gene in addition to the therapeutic gene that selectively acts only on cells that have been genetically modified and produces a toxin that kills them while not affecting the neighboring unmodified cells. Such an approach has been used in a number of human gene transfer studies aimed at the elimination of certain kinds of cancer cells, but until now the results of these studies have generally been imprecise. Nevertheless, these kinds of selective killing, or removal methods, in the event that the genetically modified cells grow out of control, or otherwise produce harm, hold promise.

3

Early Gene Transfer Experiments: Problems and Eventual Success

By the spring of 2005, approximately 700 human clinical gene transfer studies had been proposed and evaluated in the United States by local and federal oversight and regulatory bodies since they started in 1989. In addition, an ever-growing number of studies are being carried out in other countries, including United Kingdom, France, Germany, Italy, Sweden, Japan, China, Australia, and other countries. In all, thousands of research subjects have taken part in studies of cancer, the most common clinical target disease, as well as neurodegenerative diseases like Alzheimer's and Parkinson's; infectious diseases, such as HIV/AIDS; cardiovascular disease, such as coronary artery and peripheral artery insufficiency; diabetes; genetic diseases, such as immunodeficiencies, hemophilia, and metabolic liver disease; and many others. Most of the studies have been so-called phase I studies, a category of human clinical experimental work that is designed to test the safety of a new drug or method rather than its efficacy in disease treatment. But because phase I studies are generally carried out in research subjects and patients who are affected by the disease being studied, there is often the implied expectation of a clinical benefit, however, therapeutic benefit is not a formal part of the clinical experiment. A number of gene transfer studies have been designed as phase I/II studies because the gene transfer procedure has already been demonstrated to have some therapeutic effect, and in such studies the benefits of the gene transfer are to be compared to more traditional therapy in only a small number of patients. Only a very small number of studies have progressed to phase III in which the therapeutic effects are to be examined in a large number of subjects.

Advances in Genetics, Vol. 51

0065-2660/06 $35.00
DOI: 10.1016/S0065-2660(06)51003-X

In the earliest clinical phase of human gene therapy research, the conventional wisdom among investigators, the federal and other funding agencies, research and medical institutions, and patient advocacy groups, was that the technology was mature, effective, and safe and that definitive treatment of disease would be relatively straightforward and uncomplicated. For the 5–6 years from the initial clinical studies, 1989–1995, several genetic diseases were thought to be pointing to rapid clinical success, including such diseases, as cystic fibrosis, familial hypercholesterolemia, muscular dystrophy, the brain tumor called glioblastoma, genetic causes of immunodeficiency, and others. Many presentations at medical meetings and seminars seemed to be demonstrating not only high levels of safety and efficiency of the gene transfer procedures but also probable clinical benefits, even though the phase I studies were not expected to show such therapeutic effects.

When the studies were evaluated critically enough to be published, surprisingly the expected efficiency and benefits were not confirmed. In fact, for all of the early disease models, the efficiency of gene transfer and the levels of expression of the transferred genes were generally transient, very low or even undetectable, and the resulting clinical benefit nonexistent. The field as a whole received a rare public rebuke from the director of the National Institutes of Health (NIH), Dr. Harold Varmus, and his advisory committees for having permitted and even taken part in the development of an atmosphere of exaggeration and undeliverable promises. Part of the response at the oversight level was to reorganize the roles of the Food and Drug Administration (FDA) and the NIH/Recombinant DNA Advisory Committee (RAC) and to place full-regulatory responsibility at the FDA.

In 1999, another serious setback came in a gene therapy study at the University of Pennsylvania. This experiment involved subjects who had a disease of liver function called ornithine transcarbamylase (OTC) deficiency. In this disease, absence of the liver enzyme OTC interferes with protein metabolism in the liver, making it inefficient and leaving patients vulnerable to the toxic effects of incompletely metabolized protein. Some patients have very severe forms of the disease and may die even in early infancy, while others can show much milder symptoms and live relatively normal lives with dietary restriction and drug therapy.

The study at Pennsylvania involved introducing a normal copy of the OTC directly into the liver of patients in the form of a disabled adenovirus vector administered into the artery leading to the liver, a procedure that was found to be reasonably effective in animal studies in mice and even monkeys. Although some animals did show some unsettling blood changes, the feeling was that such problems could be handled well if they occurred in human subjects of the study and that the chances of benefit outweighed the potential for harm.

An 18-year-old boy named Jesse Gelsinger was suffering from one of the milder forms of the disorder, but he volunteered nevertheless, to participate in the study. Jesse was living a reasonably normal life, even with his liver disease, and the signs and symptoms of his liver problem were controlled by careful diet and drug treatments. Nevertheless, the possibility of a cure was alluring to him, and his doctors and family reported that Jesse thought he could also contribute to the well-being of other patients by participating in the experimental gene therapy study at the University of Pennsylvania. He enrolled in the research, was treated with the virus vector, and several days later he was dead, having had an explosive immune response to the virus itself. The entire medical team was shocked by this devastating and surprising outcome, and despite some clues in the preliminary animal studies pointing to potential hematological and immunological problems, no one on the medical team had imagined in their worst moments that a patient's reaction would be so severe. In retrospect, the situation was quite different. It was found only after this incident that a number of aspects of the study were inadequately designed, and the FDA was concerned enough with the entire technical and ethical design of the study to disqualify the investigators from further human clinical research and clinical gene therapy activities. The university's gene therapy institute was entirely shut down. That ban has since been lifted but the experience had an enormous impact on the field of gene therapy. It became obvious that achieving cures or even reasonably effective treatments of human disease by transferring normal copies of genes into diseased tissues would be much more difficult and complicated that most had imagined until that time.

At the time, the field of gene therapy was grappling with the implications of the Pennsylvania study, the first hints of a more convincing clinical benefit in a gene therapy study began to emerge in a gene transfer study in Paris in children suffering from severe combined immunodeficiency disease (SCID), which results from mutations of a single gene on the X-chromosome. The disease is therefore also known technically as X-SCID and has been more popularly called "bubble boy syndrome" because of the need to isolate affected children and protect them from infectious agents in the environment. The genetic deficiency interferes with the normal development of cells of the immune system, making the children vulnerable to life-threatening infections. Without the standard treatment—bone marrow transplantation—the only treatment available is the relatively ineffective therapy with antibiotics and immunoglobulin injections. A French research and clinical group headed by Alain Fischer used a traditional *ex vivo* gene transfer approach in which blood cells from these children were exposed to a retrovirus vector carrying a normal copy of their mutated gene, and the corrected cells were then returned to them by the very simple procedure of transfusion. The result was apparently a full restoration of immunological function to these children, and the children were able to start normal childhood lives—playing with the mates, going to school, and otherwise behaving like normal children. They no longer suffered from constant and lethal infections and did not require additional treatments. Similar results also began to appear in a similarly designed study carried out at the famous Great Ormond Street Hospital in London by Professor Adrian Thrasher and his colleagues. It seemed that the technical problems associated with efficient and safe gene transfer had been solved and that a disease had been conquered by gene therapy. But more dreadful surprises were in store.

In the fall of 2002, after more than 3 years of the study and after long-lasting clinical improvement had been achieved in 9 of the 10 treated patients, one of the children developed leukemia. A month or two later, a second patient developed an identical leukemia. Both responded to leukemia chemotherapy with excellent clinical remissions. But not excellent enough, and eventually the first patient died of a leukemia relapse in December, 2004. The research teams determined that the leukemia in both children was caused by the integration of the retrovirus vector into or near a known cancer-related gene in the children's blood cells, resulting in activation of the cancer gene and uncontrolled growth of the cells in the blood. The property of this class of gene transfer vectors integrating into what appear to be random sites in or near genes in the cellular DNA was known for more than two decades, and it was anticipated that the use of these vectors would result sooner or later in the disruption of important cellular functions by integrated vectors, possibly even the activation of a silent cancer-causing gene, with the possible appearance of cancer in a treated human research subject. Most people in the field of gene therapy were aware of this potential but considered it an unlikely event, certainly not a systematic problem, and that the potential for great benefit outweighed the potential for harm.

The agencies responsible for regulating and overseeing experimental human gene transfer studies in France, United Kingdom, the United States, and other countries took a careful and serious look at the problem and reached different conclusions. The French regulatory agency responsible for gene therapy studies in that country placed the Paris study on "clinical hold," meaning that no additional children could be treated until more information could become available on the reasons for the appearance of the leukemia and until the study could be reevaluated in the light of new knowledge. A similar position was taken in the United States by the advisory body at the NIH, the RAC, and by the principal American regulatory agency, the FDA, that regulates all gene transfer studies in human subjects in the United States. Several planned American studies of this type of X-SCID were, as in France, put on hold, and the RAC recommended that the studies could proceed only in children in whom the more established method of treatment by bone marrow transplantation had failed or was not feasible. At this point, approximately 16 patients had been treated in Paris and London, two had developed life-threatening leukemia, and the rest were still apparently healthy, some even more than 6 years after treatment.

The study in London was never put on clinical hold because the ethical review bodies and the regulatory agencies in United Kingdom concluded that there was more evidence for benefit in the study than the possibility of harm to participants.

In January 2005, a third child in the Paris study developed a form of leukemia very similar to that of the patients 2 years earlier. As in the first response, the French and American review and regulatory bodies restored the clinical "hold" on this approach to X-SCID, although the RAC endorsed its recommendation of 2003 that gene therapy approach should be reserved for patients in whom bone marrow transplantation was not an option or had failed. In total, 16 children were treated in the two centers, all but one of the children responded with an immunological correction lasting and a normalized childhood life—in several cases for up to 6 years, three children developed a treatment-caused leukemia, and one child died. It is a complicated confrontation between a clearly effective treatment in most children and lethal complications arising directly from the "therapy."

Another gene therapy study conducted in Italy on a different immunodeficiency disease very similar to the X-SCID study has also produced what seems to be a therapeutic benefit in several children. This disease is caused by a deficiency of a gene different from the one that causes X-SCID—in this case the gene produces the enzyme adenosine deaminase (ADA). Absence of this function causes a disease very similar to X-SCID. Although similar kinds of therapy are available as for X-SCID in addition to replacement of the enzyme itself, treatment is far from perfect. Very effective therapy has been produced lasting more than 4 years in a small number of children by virus gene transfer of the normal ADA gene into the children's blood cells by exactly the same approach as in X-SCID. Finally, in the summer of 2005, German scientists also reported impressive clinical improvement in several children suffering from yet another monogenic "inborn error of metabolism"—chronic granulomatous disease (CGD) that disables the immune system and also causes a susceptibility to life-threatening infections. Children with this disease were treated *ex vivo* with a virus vector that introduced the normal copy of the defective gene into their blood cells, just as was done with X-SCID and ADA-SCID described earlier. The children have shown a major and prolonged improvement in their immune functions and a sharp reduction in the number and severity of infections.

However, with the exception of these several immune disorders, it is fair to say that while most of the remaining 700+ gene transfer studies have proven to be generally safe, they have not led to the same kind of startling clinical response as seen in the immune deficiency diseases above. At best some of them have shown minimal or equivocal but promising clinical effects. Of course, the intent and the expectation from these kinds of early phase I clinical studies in any field of clinical research is precisely that—to test the safety of new procedures and not necessarily to demonstrate therapeutic effectiveness against the disease. Robust clinical benefit, if it appeared, would be a bonus to the primary goal of learning how to carry out gene transfer safely in human subjects and patients.

The setback with the lethal X-SCID immunodeficiency disease mentioned above was particularly unsettling because it was this study that provided the first truly credible and rigorous evidence, after 30 years of conceptual development of the field and after 15 years of human clinical experimentation, of true therapeutic benefit and perhaps even a cure of a genetic disease through gene therapy. Other than the children who developed leukemia, 13–14 of the treated children in the combined French and English studies had their immune systems completely corrected for up to 6 years after treatment by the gene transfer procedure and who, for the first time in their lives, are living normal childhood lives without the need for strict isolation and any other therapy.

Many other gene therapy studies have been and continue to be carried out all over the world. The most common disease target for these studies is cancer because of its prevalence and devastation. Cancer is a genetic disease, not usually in the sense that inborn errors of metabolism are by inheritance of cancer-producing or cancer-predisposing genes from our parents, but because the unregulated growth of cells, the hallmark of cancer, is often the result of genetic defects that we acquire by one of several ways—environmental damage, failure of normal gene repair mechanisms, and so on, after our birth in one or more of a large number of genes in our cells. These abnormal genes then facilitate the growth properties of our cells by activating growth-promoting genes or by reducing the effectiveness of the growth-reducing "brakes"—tumor ruppressor genes—that normally keep cell growth and division in check. Many of these genes are now known and more are being recognized, and the same gene transfer methods described earlier can be applied to the development of treatments for many forms of cancer. There have been many reports of some degree of improvement of human cancers, such as melanoma, head and neck cancer, colon cancer, and others, through the addition of foreign genes. For instance, many cancers of the head and neck in humans are accompanied by mutations or other deficiencies in one of the most important "brake" genes or "tumor suppressor" genes, the so-called *p53* gene. A growing number of recent studies indicate that the reintroduction of a normal copy of the *p53* gene directly into the tumors in the form of a disabled adenovirus produces more prolonged survival of patients with this distressing cancer.

Even some of our most troublesome and untreatable neurological diseases, such as Parkinson's, Alzheimer's, and Lou Gehrig's diseases, are aggressively being studied as potential targets for gene therapy. For instance, a great deal of basic research has indicated that the neurological degeneration associated with the untreatable Alzheimer's disease may be the result of deficiency of one of several so-called neural growth factors, especially one called "GDNF." Studies have begun on the reintroduction into the brain of Alzheimer's patients of a small amount of their own skin cells that have been genetically modified to enable them to make and secrete GDNF. These cells are then grafted into the most affected parts of the brain in severely affected patients, and there the grafted cells serve as small local factories for GDNF that in turn makes its way to injured cells and in theory protects them from further degeneration and death. This and similar brain gene therapy studies are in early stages of development and it is not yet clear whether they are effective or not.

The necessary conclusion from these experiences is that the techniques of gene transfer are not yet well enough understood to allow us to have much comfort in their safe and effective use in nontherapeutic settings. The corollary conclusion is that development of the technology of gene transfer and its application to disease continues to require the extensive local and national oversights that have played such an important role in bringing the field to its current state of promise. It is only through rigorous science and alert but constructive oversight that the elegant and simple concept of gene therapy will be translated into broadly delivered forms and clinical reality. It will certainly come, but a great deal of additional basic and applied medical research will be required before anyone can say with comfort that gene transfer, even for such dire diseases, is safe and effective.

4 Gene Transfer In Sports: An Opening Scenario For Genetic Enhancement of Normal "Human Traits"

Athletes have long resorted to brews and concoctions to improve their performance, and the advent of genetic technology is not likely to be overlooked by those who stand to gain so much from athletic success—the athletes themselves, their entourage. It seems very unlikely that the world of sports will remain untouched by the potential for gene-based enhancement to ensure victory in competition. In fact, sport represents one of the early and most obvious areas of human activity in which serious attempts at genetic enhancement are likely to be made, and made fairly soon. Because the tools of gene transfer are now slowly but surely being applied successfully to the treatment of life-threatening disease, there will increasingly be temptations to apply the same methods to many other human traits that represent less severe disease and are not

Advances in Genetics, Vol. 51
0065-2660/06 $35.00
DOI: 10.1016/S0065-2660(06)51004-1

disease-related at all but rather constitute normal human functions that some will wish to augment or "improve." Although, the Recombinant DNA Advisory Committee (RAC) is not receptive to nontherapeutic gene transfer in clinical human proposals, it has evaluated several studies that were submitted as potentially therapeutic but that clearly have enhancement potential. The rationale for some such eventual enhancement applications come indirectly through the therapeutic back door. For instance, one such study, evaluated in the late 1990s by the RAC, involved the sustained expression of a gene coding for the production of the hormone erythropoietin (*EPO*) for delivery through the circulation to blood-forming tissues for treatment of the anemia associated with chronic kidney disease. Another study proposed the introduction of a gene encoding the insulin-like growth factor-1 (IGF-1) into muscle undergoing degenerative changes as a result of a nerve entrapment disorder. Still another proposal involved gene transfer for the correction of erectile dysfunction. No doubt, as the techniques for safe and efficient gene transfer develop and become available in the next few years, the local and federal review and regulatory groups will see an increasing number of proposed studies involving gene transfer for conditions that further blur the line between therapy and enhancement, and even cross that boundary, such as genetic approaches to prevention of degenerative muscle and orthopedic changes in normal aging, behavior modification, memory augmentation, protection from hair loss during chemotherapy, and so on.

The move from clear therapy to more and more clearly "enhancing" gene transfer applications should come as no surprise in a society that already accepts and even seeks improvement or enhancement of many human traits through drugs and cosmetic surgery. Our society tolerates and even demands "treatment" of many conditions that may or may not represent diseases of medical deficiency but affect lifestyle issues that are important to many of us. Such manipulations include cosmetic surgery, drugs to enhance how we feel, how we interact with others, how we perform in the work place and in the bedroom, and how we and our children learn, act and look. The regulatory and oversight agencies and committees responsible for gene transfer have had a preliminary look at the potential for gene-based enhancement and for the time being have deferred the question by announcing, as the RAC has done, that it will not entertain any research proposals at the present time that are overtly aimed at an enhancement goal. The difficulty of this position is that the boundary between therapy and enhancement can be extremely tenuous and thin, and some clearly therapeutic applications could have rather immediate extension to enhancement uses. For instance, studies have been proposed to the oversight bodies that deal with improved muscle function in children with muscular dystrophy, but these exactly same methods could be used, in principle, to prevent the muscle loss that occurs as a "natural" part of the aging process. Would the prevention of age-related muscle loss be legitimate treatment or a less justifiable enhancement? Similarly, genetic tools have been proposed for the repair of joint and tissue injury in arthritis, but again, the same methods could in theory be applied to therapeutic or even preventive application in athletes to speed healing from injuries sustained in competition or training. It is even possible to envision use of the same techniques in preventive settings to enable more extensive and strenuous training or competitive effort.

Given the fact that the current oversight and regulatory mechanisms are not entertaining gene transfer proposals in nontherapeutic settings, any deliberate attempt to perform gene transfer studies or applications in any enhancement setting would necessarily mean that such manipulations in humans would circumvent accepted practice in the field and would even come into legal conflict with requirements for government approval for all new therapeutic methods. If such studies are carried out in an institution that receives any federal money for research with recombinant DNA, discovery of such surreptitious and illegal studies would make the institution vulnerable not only to legal action but also the loss of all federal research funds to the responsible investigator as well as to all investigators in all studies at the institution, which is even worse. Nothing is as dangerous or painful to a major research institution.

I. GENETIC DOPING IN SPORTS: WHAT MIGHT BE TRIED?

It is self-evident that, given the present state of the technology of gene transfer for application in therapy, it would not be possible to conduct a gene transfer application for athletic performance that would meet the requirements for ethical research in human research subjects and patients. As they would likely be performed at the present time, such applications therefore would necessarily be unethical. Why?

A gene transfer approach aimed at enhancing athletic capability, to be carried out safely and ethically, would have to conform, like all gene transfer manipulations in human subjects or patients, to the standards of ethical performance of experimental clinical research in humans outlined below in Chapter 5. Any gene transfer in sports would also have to submit to the oversight and regulatory function of a variety of local and federal bodies that oversee all human gene transfer clinical trials and that have the responsibility of evaluating the important technical, policy, and ethical issues inherent in all human clinical research. Deliberate circumvention of these procedures by investigators or institutions that receive US federal funds for recombinant DNA research could result in severe professional and legal sanctions, including loss of research funding, not only by individual investigators but also by the entire university or research institution. All established codes of human medical experimentation require that the known risks and possible adverse consequences of gene transfer studies should be fully and honestly presented to patients and research subjects, and it is only after full disclosure and informed consent that a decision can be made to proceed with the study. It is also required that the anticipated benefit outweighs the known or likely harm that can be caused to the participants. A number of local and national mechanisms have been put into place, not only in the United States but also in most other countries in which such medical science is carried out. In the United States, local institutional human subject committees and biosafety committees, as well as national oversight bodies at the National Institutes of Health (NIH) and the Food and Drug Administration (FDA) ensure that investigators are carrying out human clinical research work in compliance with these various regulations and principles. Given what we already know about drug abuse and drug-based athletic enhancement activities at all levels of sports—amateur, professional, and elite international sports—it is obvious that the principles that define acceptable human clinical research—full disclosure, informed consent, the primacy of patient and research subject safety, and autonomy of patient decision making—would not be adhered to in illicit manipulations that seek to circumvent the oversight mechanisms. Given our understanding of the science of gene transfer in humans, such studies would at best be very difficult and hazardous to the subject and therefore, unlikely for the foreseeable future to receive approval from the oversight bodies. For all of these reasons, it is impossible at the present time to perform such illicit applications ethically, safely, and honestly in ways consistent with all of the relevant tenets of human research. To do a scientifically poor and dangerous study in humans would be far less difficult, probably less successful, and certainly perilous for the athlete.

At first glance, studies of gene transfer and genetic correction for therapy might seem to be straightforward and simple—simply put a normal copy of a disease-related gene into the appropriate cells and tissue and, voila, one has a cure! But, we have learned from the many clinical studies over the past 15 years or more that how difficult it is to get it right and achieve the desired genetic and clinical results without doing harm to the subjects. Only a handful of the more than 700 clinical gene therapy studies that have already taken place under truly rigorous scientific conditions have achieved a clear therapeutic result while several have caused death and other harm to participants. It is not too complicated merely to imagine approaches to gene transfer, but carrying out an effective and safe study has proven to be excruciatingly difficult and filled with unknowns. Therefore, effective but also safe application to nontherapeutic settings such as sports might seem straightforward but is guaranteed to be frustratingly difficult. Nevertheless, the successes already achieved with immune deficiencies and imminent likely successes with other diseases establish the principle that one can alter human disease traits by genetic means and open the door widely to modification of a variety of normal human nondisease traits.

II. SEVERAL POTENTIAL SCENARIOS FOR GENE-BASED DOPING

Concepts have emerged from advances in the therapeutic applications of gene transfer of many ways in which gene-based doping might be envisioned. The approaches presented here have been discussed in many public scientific meetings and symposia and therefore, do not present novel concepts to those who would wish to abuse the current technology and apply it to enhancement purposes. One can readily imagine a number of straightforward ways in which genetic modification brought about by means identical to those used for therapy might bring about improved sports performance. These approaches are not merely theoretical. They are supported by a number of proposed human clinical studies aimed at serious disease that have been reviewed and approved by the relevant local and national agencies.

A. Muscle function

An obvious first approach to gene-based athletic enhancement might involve "improvement" of muscle function. Most competitive sport requires optimum muscle function—maximal and controlled force of contraction, optimal delivery of nutrients and optimized energy utilization in exercising muscle, and efficient removal of metabolic wastes. These properties of muscle and of blood circulation can be modified in many ways—certainly through the classic route intensive training. Scientists are coming to learn a great deal about the genes that regulate muscle function in health and in disease and even the ways in which muscle function is affected by normal kinds of training. Many of these advances are coming through the study of muscle diseases and through attempts to develop gene therapy approaches to their treatment and cure.

One of the most important model systems has been the prevention or reversal of muscle degeneration in the muscular dystrophies. Research by many muscle physiologists, such as Lee Sweeney at the University of Pennsylvania and Geoffrey Goldspink and their colleagues at King's College, London, has proven that the injection into skeletal muscle of a muscle growth factor, called insulin-like growth factor-1 (*IGF-1*), or of the gene that encodes this growth factor causes skeletal muscle to become hypertrophic. This causes the muscle to contract with greater force, recover from work more efficiently, and repair from injury more quickly. These studies were undertaken as part of efforts to develop therapies for muscular dystrophy and other degenerative diseases, and these have been extended to the degenerative muscle changes in normal aging and even to normal rats to determine the effects of this and other growth factors on muscle in normal, young animals. First, Sweeney reported that the introduction of a virus vector carrying the IGF-1 gene into the limb muscles of normal rats produced a great increase in the size and even the contractile efficiency and power of these muscles. In fact, the animals were so muscular that they came to be called "Schwartze-negger" mice. He also reported that similarly engineered rats forced to undergo weight training after having the gene encoding IGF-1 injected into their muscles showed a greater degree of enhanced muscle function than what could be achieved by weight training or IGF-1 alone and that the effects were long lasting after the weight training was halted. Such a gene therapy approach would be useful but certainly not ideal for correction of the muscular dystrophies since many muscles not accessible to direct injection, such as the diaphragm and heart muscle, would not be easily corrected by this approach. However, many investigators are developing methods for delivering genes and gene transfer virus vectors through the circulatory system to large tissues and even to entire limbs, and it seems that such methods would be feasible in humans in carefully monitored settings.

Another approach to muscle function is through the potential manipulation of a different gene that has a powerful "braking" effect on muscle growth. All genes in the body are subject to complex regulatory processes, some of which, like *IGF-1*, promote cell and tissue growth, while others act as brakes to dampen the positive stimulatory factors in order to achieve a balanced state of growth—neither too much nor too little. Myostatin is one such braking gene in muscle. It acts to counter the growth-stimulatory properties of IGF-1 and similar muscle growth factors. In an effort to improve meat production, cattle breeders managed more than 100 years ago to create two breeds of cattle, the Belgian Blue and the Piedmontese, which were markedly more muscular than normal cattle. Both breeds of cattle were found to have defects in the expression of the myostatin gene, resulting in a reduced inhibition of muscle growth and thereby an increased amount of muscle growth. Mouse geneticists then engineered a breed of mice in which the myostatin gene was inactivated and the resulting mice also had bigger, stronger muscles, similar to those of the IGF-1 "Schwarzenegger" mice of Sweeney and colleagues.

B. Erythropoietin, blood cell production, and oxygen delivery to exercising tissues

It is well recognized that erythropoietin (*EPO*) is one of the abused drugs in the enhancement of athletic performance. Erythropoietin is a normal hormone of humans and other mammals, and it is vital in the process of normal blood production. Erythropoietin is turned on under conditions in which an animal is exposed to lowered amount of oxygen in the, environment, and it acts to increase the production of the oxygen-carrying red blood cells in the bone marrow. It is one of the world's most important and widely prescribed therapeutic drugs, producing enormous benefits to patients in whom blood production is suppressed by diseases such as many kinds of cancer and chronic or end-stage kidney disease. It is a life-saving medicine. Unfortunately, athletes taking part in endurance sports, such as bicycling, realized that an *EPO*-induced rise in the level of their red blood cells also resulted in a marked improvement in their performance. The result was an uncontrolled and uninformed epidemic of *EPO* doping in the 1990s during which the allure of possible benefits of *EPO* doping was not balanced by a corresponding awareness or concern with the hazards of excessive and poorly regulated red blood cell production. Too much *EPO* can produce too many red blood cells that in turn cause red blood cell "traffic jams" in the circulation. This results into an increased tendency of blood to clot in the wrong blood vessels at the wrong time and in lethal strokes and heart attacks in otherwise perfectly healthy and fit young athletes. In order to improve the treatment of humans with life-threatening disease, clinical scientists have been examining gene-based methods for delivering the *EPO* gene so that one might have a stable source of *EPO* production in the body without the need of repeated injections of the *EPO* hormone itself. In studies reported in 2003 by scientists at Stanford University in California, the normal mouse *EPO* gene was introduced into a virus vector in a way that allowed this gene to be expressed only in the presence of a cortisone-like steroid. In the presence of the steroid, *EPO* could be produced by the vector, but in the absence of the drug, no *EPO* could be produced. This is the sort of arrangement that one would want for genes expressed in off–on ways. The vector was used to infect human skin cells that were being grown in the laboratory, and the genetically modified cells were then transplanted back to the mice—a classic "*ex vivo*" form of gene transfer. In the absence of the inducing steroid, mice containing grafts of the genetically modified cells had perfectly normal levels of *EPO* and normal levels of red blood cells in their circulation. However, when the glucocorticosteroid was applied to the grafted cells in the form of a simple topical cream, the genetically modified cells responded with an increased production of *EPO* that was, in turn, taken up by the blood circulation to the grafts and transported to the mouse bone marrow, where production of red blood cells was turned on. The level of red blood cells rose dramatically and could be maintained in the presence of the inducing glucocorticosteroid. In its absence, the level of red blood cells eventually fell slowly back to normal as the newly produced cells went through their usual lifetime of several months. Although this precise method has not yet been applied in human studies, it represents a very promising approach to serious human diseases. But in light of the role of more traditional *EPO* doping in sports, it also portends another potential avenue for doping.

Other groups studying *EPO* gene transfer have, however, demonstrated that a high percentage of monkeys, who were genetically modified to produce a foreign *EPO* gene from a virus vector, has developed life-threatening anemias, probably as a result of an immune response to vector itself. Therefore, no matter how impressive the results may be in animal testing, the gene transfer technology is not yet characterized well enough to apply in human settings in which the goal is the enhancement of normal human traits rather than the correction of a human disease.

C. Energy production and utilization

All muscle and tissue functions require carefully regulated production and utilization of metabolic energy. Muscle is a complex tissue containing a number of different kinds of muscle cells that burn energy at different rates and therefore affect muscle function and athletic performance. The slow-twitch muscle fibers are particularly fatigue-resistant, probably because of their high content of mitochondria, the power-producing elements in all cells that enable them to convert fat to energy more efficiently in contrast to the fast-twitch muscle fibers that contain fewer mitochondria and that rely completely on energy production from glucose. A group of scientists led by Ronald Evans in La Jolla, California reported that mice expressing an excessive amount of a foreign gene, peroxisome proliferator-activated receptor (PPAR delta), developed an increased number of the slow-twitch fibers. The genetically altered mice showed lower levels of intramuscular triglycerides, which are associated with insulin resistance and diabetes in obese humans. Furthermore, the mice demonstrated a reduced amount of body fat and surprisingly also became more efficient energy utilizers during endurance exercising. They were known as "marathon" mice. Because the mechanism of this effect is related to the burning of calories from fat, the genetic model is becoming important for the control and treatment of obesity, but the effects of this and probably other related genes on athletic performance are not lost on the athletic community.

To do a rigorous and safe clinical trial in the muscular dystrophies, in disorders of blood production, in many diseases involving defects in metabolism and in many other human illnesses is difficult and dangerous enough but is often scientifically and ethically accepted in the name of treatment of life-threatening disease. Similarly, studies aimed at the correction of the life-threatening anemias that accompany malignancies or kidney disease through the introduction of the gene encoding the hormone *EPO* that regulates red blood cell production are extremely complex and plagued by the consequences of inadequately or inappropriately controlled blood production—strokes and other cardiovascular catastrophes. For instance, the severity and the incompletely understood nature of such adverse results have been underscored by the demonstration of two independent research groups of lethal anemias resulting from the expression of a foreign *EPO* gene in monkeys.

By definition, any genetic manipulation undertaken with current technology for the purpose of enhancing sports performance would be undertaken without the required degree of proven safety, full disclosure and informed consent, and necessary oversight and regulation to ensure protection of the subjects (see Chapter 5) and, therefore, would be foolhardy, unethical, and dangerous.

III. GENETIC ENHANCEMENT

These kinds of studies would obviously represent a departure from gene therapy as it has come to be recognized as an approach to preventing or correcting disease. In all countries in which gene therapy is being pursued under the all the required review and regulation, all gene transfer studies carried out until now have been aimed at conditions that would be universally seen as diseases—cancer, neurodegenerative diseases such as Parkinson's and Alzheimer's diseases, cystic fibrosis, immunodeficiency states, etc. In the potential studies described previously, the intent is not to cure a disease but rather to make a normal human trait supranormal even "better" than normal. This potential was recognized in the earliest days when the concept of gene therapy was first emerging. It was evident that, once the gene transfer technology became feasible, attempts would eventually be made to transfer foreign genes to modify traits that represented human properties somewhere between normality and disease and then eventually to completely normal traits. For instance, the loss of muscle mass and function as a part of human aging can be considered to be a completely normal phenomenon and therefore accepted with no need for "treatment." On the other hand, some could maintain with good reason that genetic methods to improve muscle function, as developed in studies of terrible muscle diseases such as muscular dystrophy, would increase quality of life for millions of people in the same way that treatment of other aspect of human aging—arthritis, cancer, memory loss, are all acceptable parts of modern medicine. It would not be a great conceptual or technical leap then to apply the same technologies to completely normal human traits such as muscle function or blood production in normal young athletes for the specific purpose of improving athletic prowess. In that way, the world of sports may be one of the provinces of human activity to become an early target for the application of existing and future advances in the field of human gene therapy for the enhancement of nondisease human traits. A number of comprehensive moral and ethical analyses of the issues of gene-based human "enhancement" have been undertaken by individual scientists and ethicists as well as a variety of the most authoritative and thoughtful deliberative groups, including the American Association for the Advancement of Science, the Hastings Center, the most eminent bioethics think tank in the United States, and by the US President's Council on Bioethics in their report, "Beyond Therapy" (www.bioethics.gov). Although some of these deliberations have concluded that most kinds of genetic modification aimed at enhancing nondisease human traits, such as intelligence, and should be pursued with all deliberate speed, desirable personality traits, etc. are acceptable advances of science, most of the more thoughtful discussions have concluded that there are severe moral objections to such efforts, even if and when the technical limitations that we now face are overcome and such changes might be produced safely without harm to the subject or, for that matter, to future generations. These have also forced a new look at the role of a sport in a society, the very concept of a sport, and its apparent evolution from a rule-based noble and romantic striving for individual achievement to potentially grotesque uncontrolled biotechnology.

IV. A SPECIAL CASE SOMEWHERE BETWEEN GENETIC ENHANCEMENT AND GENETIC THERAPY IN SPORT: TREATMENT OF DISEASE AND REPAIR OF INJURY IN AN ATHLETE

Sport presents an interesting and possibly unique quandary that illustrates the fuzziness of the differentiation of gene therapy from genetic enhancement. It is inevitable that effective gene-based methods will become available and possibly even routine for treatment of some diseases or the repair of injuries in athletes, both human as well as nonhuman. Athletes of course frequently suffer injuries that threaten their health not only in the same way that injury threatens us all, but also that threaten their very livelihood and even the collective and personal sport objectives with considerable financial impact. Athletes need and deserve access to the best medical procedures and methods to repair their injured muscle, tendons, bones and all other tissues just as all injured patients need and deserve the best in medical repair technology. Athletes must not be deprived of the best medicine available to all others. However, it seems very likely that some methods for treatment of injuries or illness in athletes result inadvertently not only in restoration of the original normal function but that produce an enhanced physical function—stronger muscles, ligaments, and tendons less vulnerable to further injury, and so on. That is probably just as true of currently available methods of injury repair, but enhancing physical function is usually not a significant goal in treatment of injuries and illness by existing methods. Well-treated athletes are not excluded from competition because previous therapy for an injury or an illness has left them with possibly enhanced athletic capability. However, it is now becoming possible that enhanced function might not be merely an unintended consequence of treatment but in fact may become a part of the goal of the "treatment."

For instance, it has been demonstrated in animal studies that treatment of some joint and musculoskeletal injuries and illness with growth factors such as IGF-1 markedly increases the degree and speed of tissue repair. If similar gene-based techniques were also to be proven to be effective in humans, it would seem to be of great benefit to offer such therapies to all injured people, including injured athletes. Could athletes ethically be deprived of such essential medical procedures? Of course not. But such an athlete's competitors might be placed in a position of unfair disadvantage. Even more troublesome, of course, is the possibility that prophylactic treatment of an injury-prone muscle in an athlete with such a factor for the direct purpose of providing such the benefit of improved athletic capability could readily become feasible. Such an application could certainly be consistent with the best principles of preventive medicine but would certainly cause discomfort in the sports context. And by sport we should not restrict ourselves only to human athletes. Race horses and other animal athletes could just as easily become the subjects for such therapy-enhancement applications.

5 Ethics and Oversight in Clinical Trials: Attempts at Gene Doping Would Not Conform to Accepted Ethical Standards

We present here a description of acceptable standards of human experimentation as set forth in many codes that govern modern medical research world wide. These principles and standards should be kept in mind as one ponders the questions related to gene doping—would the illicit, scientifically premature, and furtive use of gene transfer methods for enhancing athletic performance conform to all, or even any, of these principles?

I. THE ETHICS OF RESEARCH WITH HUMAN SUBJECTS

All gene transfer studies in human subjects or patients for the purpose of preventing or treating even the most fearful and dire disease are highly

Advances in Genetics, Vol. 51
0065-2660/06 $35.00
DOI: 10.1016/S0065-2660(06)51005-3

experimental. As we have seen in the cases of immune deficiency diseases in earlier discussion, there are many ways in which such procedures can go seriously wrong, but we all—patients, doctors, and society—all accept such methods in the cause of easing suffering. All progresses in medicine require multiple layers of research and testing—some in the laboratory in model systems, some with test animals, and some necessarily with human subjects. Each layer of research and testing requires a set of appropriate ethical standards for how such research should be conducted—standards for chemicals and cells in the laboratory are not relevant to those for animal research, and these standards will not be identical to those for human test subjects, human remains, cadavers, tissues, biological fluids, embryos, or fetuses. Research, at all of these levels, is appropriate and necessary, and there is nothing intrinsically wrong or unethical in experimentation with human beings, as long as such studies conform to socially approved and agreed-upon standards. Without human clinical research studies, we would have no modern therapies for cancer, heart disease, degenerative neurological disease, diabetes, and other scourges of our society. It is therefore no surprise that the medical and scientific communities in most countries in which medical research is carried out adhere to the model of ethics review that has emerged in the international community in recent decades. This model generally involves the application of international norms by many layers of local and national review and regulation by multidisciplinary and independent committees and boards that review the ethical standards of all research studies with human beings.

At this stage of development of gene transfer technology in human beings, human gene transfer and gene therapy studies are highly experimental and, therefore, must be subject to the accepted norms and standards that pertain to all human experimentation. As is true in all other areas of human clinical research and experimentation, failure to comply with these, or other comparable universally accepted standards, must rightly be considered medical malpractice and professional misconduct when carried out by licensed professionals or even criminal in cases in which the use of drugs and other materials or devices is not in accordance with governmental licensing and commercial requirements. It is reasonable to keep in mind the following question regarding possible gene transfer applications in athletes to enhance performance—would they conform to such standard requirements for human research if carried out openly?

Human experimentation is one of the bedrocks of progress in modern medicine, and a large number of our most valuable modern medical techniques and tools have relied to a great extent at one point or another of their development on clinical experiments with human beings. The ability of modern medicine to save lives and ease an enormous amount of suffering, treat cancer, transplant organs, perform innovative surgical procedures, and develop new drugs have required long and at times unsure periods of risky and even dangerous experimental studies, during which harm has befallen some study participants because of the inability of medical science to anticipate and prevent all of the adverse events inherent in new and unproven technology. We are all, however, as potential patients reaping the rewards of those periods of potentially dangerous experimentation and uncertainty. To recognize these hazards and protect research subjects and desperate patients from the harms of inappropriate experimentation, many ethical codes of human experimentation have arisen, particularly in the latter half of the twentieth century, spurred by the atrocities and unethical human experimentations that were carried out during World War II.

Following World War II, the most influential of these codes was the set of principles that were enunciated as part of the Nuremberg War Crimes Doctors' trial in 1947. This code identified the following 10 conditions that must be satisfied in order for any experiment with human subjects to be ethically acceptable. The following principles were enunciated by the Nuremberg tribunal:

The great weight of the evidence before us to effect that certain types of medical experiments on human beings, when kept within reasonably well-defined bounds, conform to the ethics of the medical profession generally. The protagonists of the practice of human experimentation justify their views on the basis that such experiments yield results for the good of society that are unprocurable by other methods or means of study. All agree, however, that certain basic principles must be observed in order to satisfy moral, ethical and legal concepts:

1. *The voluntary consent of the human subject is absolutely essential. This means that the person involved should have legal capacity to give consent; should be so situated as to be able to exercise free power of choice, without the intervention of any element of force, fraud, deceit, duress, overreaching, or other ulterior form of constraint or coercion; and should have sufficient knowledge and comprehension of the elements of the subject matter involved as to enable him to make an understanding and enlightened decision. This latter element requires that before the acceptance of an affirmative decision by the experimental subject there should be made known to him the nature, duration, and purpose of the experiment; the method and means by which it is to be conducted; all inconveniences and hazards reasonably to be expected; and the effects upon his health or person which may possibly come from his participation in the experiment. The duty and responsibility for ascertaining the quality of the consent rests upon each individual who initiates, directs, or engages in the experiment. It is a personal duty and responsibility which may not be delegated to another with impunity.*
2. *The experiment should be such as to yield fruitful results for the good of society, unprocurable by other methods or means of study, and not random and unnecessary in nature.*
3. *The experiment should be so designed and based on the results of animal experimentation and a knowledge of the natural history of the disease or other problem under study that the anticipated results justify the performance of the experiment.*
4. *The experiment should be so conducted as to avoid all unnecessary physical and mental suffering and injury.*
5. *No experiment should be conducted where there is an a priori reason to believe that death or disabling injury will occur; except, perhaps, in those experiments where the experimental physicians also serve as subjects.*
6. *The degree of risk to be taken should never exceed that determined by the humanitarian importance of the problem to be solved by the experiment.*
7. *Proper preparations should be made and adequate facilities provided to protect the experimental subject against even remote possibilities of injury, disability or death.*

8. *The experiment should be conducted only by scientifically qualified persons. The highest degree of skill and care should be required through all stages of the experiment of those who conduct or engage in the experiment.*
9. *During the course of the experiment the human subject should be at liberty to bring the experiment to an end if he has reached the physical or mental state where continuation of the experiment seems to him to be impossible.*
10. *During the course of the experiment the scientist in charge must be prepared to terminate the experiment at any stage, if he has probable cause to believe, in the exercise of the good faith, superior skill and careful judgment required of him, that a continuation of the experiment is likely to result in injury, disability, or death to the experimental subject.*

At the heart of the code are the principles that participation in experimental studies must be voluntary based on informed consent and justified by a potential for benefit that outweighs the potential for harm. Being "informed" requires that the study participants understand what will be done and agree to bear the expected risks and reap the anticipated benefits (see later). These principles have served as the basis for many variations and subsequent modifications by worldwide medical and scientific bodies, including the influential codes developed by the World Medical Association in the form of its Declarations of Helsinki—first stated in 1964, and amended and extended many times, most recently in 2004. These declarations have provided a valuable extension to the Nuremberg Code by providing guidance for human experimentation, not only for therapeutic research but also for studies that involve not only potential new therapies but also studies that are intended to acquire new knowledge and potentially to apply to other nontherapeutic uses. Parts of the Helsinki Declaration as revised in 2004 that are relevant to the potential inclusion of healthy, young subjects in nontherapeutic research state the following:

It is the duty of the physician in medical research to protect the life, health, privacy, and dignity of the human subject.

1. *Medical research involving human subjects must conform to generally accepted scientific principles, be based on a thorough knowledge of the scientific literature, other relevant sources of information, and on adequate laboratory and, where appropriate, animal experimentation.*
2. *Appropriate caution must be exercised in the conduct of research which may affect the environment, and the welfare of animals used for research must be respected.*
3. *The design and performance of each experimental procedure involving human subjects should be clearly formulated in an experimental protocol. This protocol should be submitted for consideration, comment, guidance, and where appropriate, approval to a specially appointed ethical review committee, which must be independent of the investigator, the sponsor or any other kind of undue influence. This independent committee should be in conformity with the laws and regulations of the country in which the research experiment is performed. The committee has the right to monitor ongoing trials. The researcher has the obligation to provide monitoring information to the committee, especially any serious adverse events. The researcher should also submit to the committee, for review, information regarding funding, sponsors, institutional affiliations, other potential conflicts of interest and incentives for subjects.*
4. *The research protocol should always contain a statement of the ethical considerations involved and should indicate that there is compliance with the principles enunciated in this Declaration.*
5. *Medical research involving human subjects should be conducted only by scientifically qualified persons and under the supervision of a clinically competent medical person. The responsibility for the human subject must always rest with a medically qualified person and never rest on the subject of the research, even though the subject has given consent.*
6. *Every medical research project involving human subjects should be preceded by careful assessment of predictable risks and burdens in comparison with foreseeable benefits to the subject or to others. This does not preclude the participation of healthy volunteers in medical research. The design of all studies should be publicly available.*

7. *Physicians should abstain from engaging in research projects involving human subjects unless they are confident that the risks involved have been adequately assessed and can be satisfactorily managed. Physicians should cease any investigation if the risks are found to outweigh the potential benefits or if there is conclusive proof of positive and beneficial results.*
8. *Medical research involving human subjects should only be conducted if the importance of the objective outweighs the inherent risks and burdens to the subject. This is especially important when the human subjects are healthy volunteers.*
9. *Medical research is only justified if there is a reasonable likelihood that the populations in which the research is carried out stand to benefit from the results of the research.*
10. *The subjects must be volunteers and informed participants in the research project.*
11. *In any research on human beings, each potential subject must be adequately informed of the aims, methods, sources of funding, any possible conflicts of interest, institutional affiliations of the researcher, the anticipated benefits and potential risks of the study, and the discomfort it may entail. The subject should be informed of the right to abstain from participation in the study or to withdraw consent to participate at any time without reprisal. After ensuring that the subject has understood the information, the physician should then obtain the subject's freely given informed consent, preferably in writing. If the consent cannot be obtained in writing, the nonwritten consent must be formally documented and witnessed.*

In summary, all experimental procedures in human patients or normal volunteers must conform to these standards or to one of many other codes that have been derived from these original sets of principles. The scientific and ethical justification of clinical studies with humans must be based on an evaluation of the following major questions: (1) Is the scientific basis for the study appropriate? (2) Does the importance of the study justify any potential risks? (3) Are the potential risks and benefits honestly presented to the participants? (4) Is the safety of participants adequately protected? (5) Do the participants have sufficient understanding (i.e., to consent knowingly) to their participation in the study? (6) Are the investigators driven by any conflicts of interest that compromise the well being of study participants? (7) Is the proposed manipulation with human subjects an open, honest, and well-designed "study," or is it a self-interest exercise in deception?

As an answer to the question posed at the beginning of the chapter, the illicit, scientifically premature, and secretive use of gene transfer methods for enhancing athletic performance conform to all, or even any, would not come even close to conforming to these sets of ethical principles. The procedure would therefore be unethical.

II. ADDITIONAL LEVELS OF REGULATION AND OVERSIGHT IN GENE THERAPY RESEARCH

Another test of the appropriateness of gene transfer application in athletics would be whether such work is carried out under the accepted current standards of regulation and oversight to which all legitimate gene transfer studies with human subjects and patients are subjected. From the conceptual beginnings of the field of human gene therapy in the early 1970s and the onset of human clinical trials in 1989–1990, it was obvious that this field of clinical research was going to require, at least in its early stages, an unusually high level of oversight and regulation.

In the United States, the fields of gene transfer research and gene therapy have emerged in the context of not only these codes of human experimentation but also as part of an extensive system of local and national oversight and regulatory mechanisms to ensure maximal protection for participants in gene transfer research studies. An investigator, who wishes to undertake any human gene transfer study in research subjects or patients, must describe the proposed study to, and obtain approval from, a number of committees and agencies that include those in the next sections.

A. US and Canada: Institutional Review Boards (IRBs), Institutional Biosafety Committees (IBCs) and Research Ethics Review Boards (REBs)

These committees have been established under federal mandate at all major academic and commercial research institutions and are responsible to ensure the safety of people being recruited into clinical research studies. The committees are responsible for determining if a proposed study with human subjects conform to the standards of acceptable research as enunciated by the many appropriate codes of clinical research stemming from the Nuremberg and Helsinki codes and as modified and extended by many subsequent sets of standards from national medical associations and professional societies in many countries, the National Institutes of Health in the U.S., and many others. They are responsible for evaluating the scientific validity of a study, its safety, compliance with standards of informed and voluntary consent, and then for making recommendations to the investigators and responsible institutional officials on whether a study should proceed. The members of such committees must represent a broad range of scientific and social expertise at the institution at which the proposed study will be carried out, including at least one noninstitutional member of the public. For investigators at institutions that may not have established their own Institutional Review Boards (IRBs), a number of free-standing IRBs are available for contract review of gene transfer proposals. In the IRB review of a gene transfer study, investigators are asked to describe the scientific background of their proposal, present supporting information obtained in preliminary laboratory studies, and explain why it is important to extend the study to human beings. They are also required to describe the procedures that they have instituted for the protection of the subjects, addressing all of the issues raised by the Nuremberg and Helsinki codes and the many other accepted codes of conduct in medical research. Is the study based on the soundest possible science, is there a clear description of the risks and benefits, is the balance in favor of the benefits of the study, do the subjects have enough clearly presented information to understand what will be done to them, and do the subjects give their free consent?

In addition to the IRB committees, institutions are required to have an Institutional Biosafety Committee (IBC) mandated by the federal government to evaluate the also review the proposed study and determine the safety issues pertinent to the human application.

In Canada, similar bodies called Research Ethics Review Boards (REB) have been established to help ensure that ethical principles are applied to research involving human subjects. As in the case of an IRB, an REB has the responsibility for independent, multidisciplinary review of the ethics of research to determine whether the research should be permitted to start or to continue. The REB structure specifies that, while it is essential that members of the REB have a general expertise in the relevant sciences or research disciplines, it is also important that its membership includes representatives knowledgeable in medical research ethics and in legal matters.

B. National review and regulation in the US: The FDA and the NIH/RAC

In the case of gene transfer research in the United States, investigators are also required to obtain permission at the national level by obtaining an Investigation New Drug (IND) approval from the Food and Drug Administration (FDA). Approval at the level of the FDA requires extremely rigorous study design and the presentation of evidence that the investigators have considered and maximally protected the safety of study participants. At the same time, before the local IRB allows the recruitment of any research subjects in the study, investigators, working at any institution that receives US government grant for recombinant DNA research, must also submit the proposed study to the Recombinant DNA Advisory Committee (RAC) and the federal National Institutes of Health for review. Like the FDA, the RAC evaluates the scientific basis for the study and research subject protections, and it also notifies the investigators and their institutions regarding the concerns that should be addressed before the investigators may proceed with their research. Although the RAC is not a formal regulatory body, a clinical study may not proceed unless the RAC has a chance to review it and notify the investigators, the sponsoring institutions, and the FDA of its concerns. Its recommendations are of great influence on the FDA deliberations and on the decisions of the local IRB and IBC committees.

This elaborate multitiered mechanism of review and regulation for all gene transfer studies in human beings reflects the strong common understanding that these gene manipulation studies are highly experimental, full of real and potential dangers to the participants, and have not yet reached the point where they can be considered safe and reliable enough to use outside the treatment of serious human disease.

As a reflection of the understanding that the methods of gene transfer are not scientifically ready for genetic enhancement in general and sport doping in particular, it is important to realize that the federal guidelines that define the function of the RAC state clearly that the RAC will not entertain a review of any proposed research study that is aimed at modifying normal traits for the purpose of genetic enhancement. Without doubt that constraint is likely to come under continuing re-evaluation in the future as the boundary between therapy and enhancement becomes ever more indistinct. The tools of gene therapy are the demand for "treatment" of human traits that lie somewhere between normality and disease will increase as our molecular and genetic understanding of these conditions grows. The example of dwindling muscle strength in the "normal" human aging process has been mentioned earlier, and it is inevitable that completely safe methods for the genetic correction of muscle diseases such as muscular dystrophy will be applied to aging subjects to preserve muscle function into the latter years to provide an improved quality of life. How many of us would reject such an offer? One might make similar arguments for other conditions such as loss of memory and cognitive power in aging. But it is obvious that the tools and techniques for gene transfer are not yet well enough understood to move soon into enhancement applications. One might readily imagine quite inappropriate and frivolous applications—striving without obvious limits for ever bigger, ever better, ever stronger, and ever smarter. To what end?

III. THE ETHICAL PRINCIPLES USED BY REVIEW BODIES

Given the understanding that some experimental studies in human subjects and patients are an acceptable part of acquiring knowledge of normal human biology, all the review and regulatory bodies involved in gene transfer studies are charged with two principle tasks common to all human experimentation. The first is to minimize risks of harm to research subjects and to assure that the risks are reasonable in relation to the expected benefits to the subject or in relation to the importance of the knowledge likely to be derived. Harm could include physical harm such as pain and serious injury, psychological damage such as stress or guilt, invasion of privacy, and breach of confidentiality. Risk varies from minimal to severe, with minimal risk generally taken to mean a risk of harm no greater than a subject would encounter in daily life or during routine performance of physical or psychological testing. An undeliverable or excessive promise of reward or benefit can be considered another form of harm—a particularly appropriate issue in the case of healthy young athletes in the context of genetic enhancement applications. Identification of benefits or potential benefits is also a vital part of the decision process in these panels. Benefits include potential increased well-being of the subjects, but since most gene therapy studies to date have been phase I studies in which the goal is merely to establish safety and not efficacy, research participants cannot expect to derive any direct benefit. In that case, the true benefits fall on others—on the society in general through acquisition of new knowledge and of course potentially to future patients. It is true that is some fortunate cases, as in the immunodeficiency diseases reviewed in Chapter 3, a study intended to be essentially a phase I study proved of great benefit to the subjects being studied. Most investigators are usually no so lucky in phase I studies.

A decision by review and regulatory bodies to proceed or not then comes down to an analysis of both the risks and the benefits of a study and the estimation of a so-called "risk–benefit ratio"—is there greater benefit than harm and is the harm justified? In the case of unproven experimental procedures in normal, healthy young athletes, most or all risks of potential harm would almost certainly rise well above the threshold of minimal risk. If such a procedure were to be carried out in accordance with all of the accepted ethical standards, knowledge of those risks and the expected benefits to an individual should have been derived from appropriate preclinical studies.

The decision to enter into an experimental procedure requires full disclosure by the investigator, informed and voluntary consent by the subject, with the word *informed* underscored in the context of potential genetic doping. There are some basic requirements that must be satisfied when seeking informed consent from research subjects. It should be clearly stated that the subject is being invited to participate in a research project. There should be a comprehensible description of reasonably foreseeable harms—physical, psychological, social—and benefits that may arise from participation. Assurance that prospective subjects are free to decline their participation must be provided and that subjects have the right to withdraw at any time without prejudice. The possibility of commercialization of research findings and the presence of any actual or potential conflicts of interest on the part of researchers, their institutions, or sponsors.

Sadly, in the murky world of sport doping, it is unlikely that any or all of these basic requirements of experimental studies with human subjects would be satisfied. The risks would be hidden, the benefits exaggerated and the risk/benefit ratio merely a guess. Under those conditions, the subject—the athlete—could hardly be expected to give informed and voluntary consent. The procedure would be unethical.

6

International Cooperation and Regulation: The Banbury Workshop (2002)

I. INTERNATIONAL REGULATION OF SPORT APPLICATIONS OF GENETIC TECHNOLOGY

The growing concern that gene transfer technology was becoming a potential threat to sports led to a workshop sponsored by the World Anti-Doping Agency (WADA) at the Banbury Center of the Cold Spring Harbor Laboratory on Long Island, New York, in March 2002 to catalyze a discussion among the disparate communities who had not previously come together to identify and discuss the issues posed by the impending possibility of gene-based doping in sports. The WADA Health, Medicine and Research Committee, under the chairmanship of Professor Arne Ljungqvist of Sweden, brought together the communities concerned with these questions—the International Olympic Committee

Advances in Genetics, Vol. 51

0065-2660/06 $35.00
DOI: 10.1016/S0065-2660(06)51006-5

(IOC), the international sports world, gene transfer scientists and geneticists, ethicists and public policy experts, and lawyers. The goal of the meeting was to identify the nature of the challenges to sports posed by gene transfer technologies and to provide a mechanism through which the details and implications of scientific progress in human gene therapy and the growing pressures in sports for illicit use of gene transfer technology would be brought to the attention of the athletics world in a realistic and unexaggerated way. The World Anti-Doping Agency hoped that such a setting would serve to develop a common understanding of the problem and develop a shared scientific and sports language that could develop mechanisms to shape the issue rather than merely to respond to the inevitable future disasters and misadventures. In fact, the Banbury meeting was the first such effort organized to bring together the "stakeholders" and interested parties—athletes, athletic organizations, scientists, WADA, and other oversight bodies—to develop a common understanding and a common language for this intersection of science and sports.

The format of the meeting included scientific summaries of the general principles of gene transfer, the scientific approaches to gene transfer in general and the status of clinical gene therapy applications, as well as the legal, ethical, and public policy aspects of gene transfer for the enhancement of sports-related human traits. The workshop began with the following introductory keynote address by WADA Chairman Richard Pound.

A. Sport: where talent and genetic manipulation collide

To begin, I want to thank all the participants in this workshop for taking the time to make their expertise and experience available on this occasion. This is the first time that a conference of this nature has been held at this level, with the presence of the sports movement, scientific experts, medical ethicists and a unique international agency, such as WADA, to consider the implications of genetic manipulation as they may relate to society in general, and to sport, as a subset of society, in the near and longer term future. Speaking from the perspective of sport, I am delighted that everyone has agreed to contribute and I hope that we will develop some useful conclusions and recommendations during the course of our time here. I can barely imagine the intellectual excitement of those of you who are scientists as the prospect of mapping the human genome and then working with it became a realistic possibility, thanks to advances in science and the enormous power of computers that enabled you to process millions and billions of items of data at speeds that were unimaginable only short years ago. As a scientific achievement, this may well rank among the most important in our history to date, and I am sure you share an enormous amount of pride and satisfaction in the work you have accomplished—pride and satisfaction that you certainly deserve. Identification of specific genes and their relation to the human condition as we know it at the beginning of the new millennium presents opportunities, heretofore impossible, to fight against disease and to improve the physical, and perhaps mental, health of the world's population. The panorama that has now been opened is unparalleled. Your principal problem, faced with these scientific results, may well be with the impatience of a world population that seeks instant cures for diseases that can now attract only palliative care, not cure, but that may now lead even to eradication of certain diseases or conditions in the future. As members of society in general, we share the excitement and the impatience of which I speak. As members of society in general, we also know that we know very little about what may become possible. What now appears possible, and even likely, is virtually unknown to the uninitiated. The prospect of miracles shimmers in front of everyone, the sick, the lame, the halt, and the blind. Will we live to the age of 150 years? Will senility and Alzheimer's disease, even heart disease, become distant memories? Will cancer become like small pox, something that our descendants will read about as an unknown scourge of a distant past. Will the great physicians of our day be seen as the primitive painters on the medical walls of caves? Like the leechers of the past, whose so-called skills are exposed as scientific charlatanism, they being totally ignorant of the real causes of disease? Will we shake our heads at the shortcomings of what was considered, even in our time, to be the leading edge of science? Will we describe the physicians of our time as little short of witch doctors, mumbling chants over smoke and mystery, hiding their evident ignorance of scientific fact behind an apparent omniscience, impenetrable to the uninformed laity? Will we ask, who were these healers, who apparently knew naught of which they spoke and whose wisdom we dared not challenge? All this is worrying enough, but we know even less about those who will control the directions in which the new knowledge will be applied. Nor do we know how this knowledge will be applied and for what purposes. Although we have all been conditioned to the notion that knowledge per se is neither good nor evil, but simply fact, neutral of and by itself, we have also experienced the frailties of human nature. While the vast majority of people in control of facts and knowledge use both for laudable purposes, it cannot be denied that there are others willing to subvert such a position for baser purposes. We do not know the scientific, moral, ethical, or legal constraints that will apply in respect of this new knowledge. Wearing, however, our hats as representatives of sport, as practiced at the highest levels, I must say that the field of genetics and the concept of genetic manipulation are subjects that are potentially very troublesome to us. We have lived through the eras of improved training methods, improved techniques, improved equipment, and the evolution of sport from a leisure activity for the diversion of a few wealthy men to a universal phenomenon that has become little short of a basic human right. These have been, for the most part, positive advances within society. Working classes now participate. Women in almost all parts of the world are free to participate and compete in sport. Participants benefit from safer conditions and equipment that will reduce the prospect of serious injury. Sports medicine has developed as a separate medical specialty, to help avoid and to cure injuries that may be suffered.

But we have also seen science and medicine subverted within the practice of sport, especially in matters of doping. Athletes have used substances that are dangerous to their health, in order to get a short-term improvement in their performance. They have done this on their own initiative, but have also, in far too many cases, been encouraged, even forced to use the same substances. Such pressures have come from doctors, trainers, sports officials, and, in the worst of cases, from governments. Measures have been instituted within the sports community to reduce and eliminate such practices. Rules have been established to protect the health of athletes and to preserve the fundamental ethical basis of sport. Sanctions are imposed by sports authorities in cases of breaches of the rules. Tests, in-competition and out-of-competition, some with no advance notice, are imposed. Some governments have enacted legislation to prevent the use of certain substances. All these measures have been only partially successful. Doping in sport continues. There are anecdotal suggestions that drug use in sports is, in fact, increasing. For those of you unfamiliar with

WADA, I should perhaps take a moment to describe it. WADA was created at the end of 1999 as a partnership between the public authorities of all five continents and the Olympic Movement. Its governance structure is a Foundation Board consisting of equal representation of the public authorities and of the Olympic Movement. It was formed to coordinate a worldwide fight against doping in sport. It has become painfully obvious that neither the sports movement nor the public authorities, acting alone, can solve the many issues; only a collective response can have any hope of success. WADA represents the most ambitious and concerted response to date in the fight against doping in sport and, although still in its early stages, indications are that significant progress has already been made. Its strategic plan is also ambitious, across the fields of harmonization, legislation, research, testing, and education. Its research budget is in excess of $30 million for the next 5 years, focusing on oxygen-carrying agents, growth agents and the usual suspects of anabolic steroids, and others. To focus on education, one of the main areas of activity will be that of education. A renewed commitment to health and ethical principles will be the best solution to the problem. Only a fundamental attitudinal change can bring about a complete solution. Sport is a humanistic endeavor, based on the mutual consent of the participants to abide by certain rules and, within those rules, to see how far the natural talents and abilities of the participants can take them, measured against the natural talents and abilities of other participants. It is meant to be, and should be, fun. It is based upon some fundamental ethical values, which hold that the rules, agreed upon by the participants, will be followed. In some cases, however, the rules are broken by those who do not share the values that they, ostensibly, espouse. One area involves the use of performance-enhancing drugs that are prohibited by the rules of play. These are used by those who cheat and who deprive sport of its fundamental validity. The challenge for sport is to maintain its ethical integrity and to expose and to remove the cheaters who seek to destroy sport as it has been conceived and practiced by the great majority of participants. Testing and sanctions will still be required as a protective measure and methods of detection will need to be perfected and kept up-to-date, but, in the long run, it will be the ethical platform upon which sport rests that must become paramount. Sport is artificial, to the extent that it involves all participants accepting and abiding by rules upon which they agree from the outset. A participant who breaks those rules to gain an unfair advantage "steals" from his fellow competitors, whether *in monetary terms or simply in the integrity and the glory of the outcome of the competition. That is unethical and should be out of the question in the first place and condemned should it nevertheless occur. Our source of concern with respect to genetics is based, to some degree, on our experience with doping in sport. I want here to throw some challenges in the direction of the medical and scientific professions. We have been singularly unsuccessful in getting consensus amongst medical practitioners, in particular, to condemn the use of drugs in sport, where the usage is not therapeutic but solely for performance enhancement. The medical profession hides, in my respectful view, behind the shibboleth of Patient Autonomy and, in some particularly sophistic cases, behind the self-justification that the athletes might well use the prohibited substances anyway, so it is better that they do so under medical supervision. Let me give a domestic example to prove the point. In 1988, the Canadian sprinter, Ben Johnson, tested positive following his Olympic victory in the 100 m at Seoul. He had set a world and Olympic record in the event but was found to have had the metabolites of the anabolic steroid, stanozolol, in his system. He was disqualified and his medal withdrawn. A few months later, I participated at a conference, in which there was a panel of medical ethicists present, discussing the issue of doping in sport. At the end of the panel discussion, I put a question to the panel. I said that I wanted to ask a hypothetical question, to be answered by the ethicists. I said that they should assume they had a patient who came to them and said, "Hi, my name is 1/4 let's pick one at random 1/4 Ben Johnson. I am a pretty good runner and I am training for the Olympics. I do not, however, think that I can win unless I use anabolic steroids, which, I want to inform you, are completely illegal in the Olympics. They have no therapeutic value at all and I have no medical condition that they will address. They will simply help me to run faster, even though they are prohibited, but I nevertheless want you to prescribe them for me. "In such circumstances, I asked, what would you, as medical practitioners and ethicists, do to a person, and to my astonishment, each of them said they would prescribe the drugs. On what possible basis, I asked? "Autonomy of the patient," they intoned. I realized then that the medical profession had abandoned any pretence of ethical engagement in sport and that, instead of being of assistance in the ethical involvement on an important dimension of social behavior, it was as likely to assist cheaters as it was to help keep competition pure. I confess to have found this attitude as disappointing as it was inexplicable. It is one reason why we want to have this occasion at the Banbury Center as we face the prospect of genetic manipulation which will probably make drugs like stanozolol look like the dark ages.*

Patient autonomy, surely, cannot be the complete answer. There is, at the very least, some kind of fiduciary duty on the part of this learned profession, to live by the full implications of the Hippocratic Oath, to which they all subscribe, to do no harm. Choosing a "treatment" or allowing a patient to choose a treatment that does obvious harm is, in my respectful view, a breach of that Oath. There is, in my view, no basis for a medical doctor to be unaware that a patient under his or her charge is using prohibited substances for nontherapeutic purposes. There are too many cases in which cheating in sport has been made possible with the active help, and even encouragement, of medical practitioners. I do not want to suggest that the medical profession is devoid of ethics. I am sure that many professional associations have developed ethical rules regarding appropriate actions in this field, as well as other organizations, such as the World Medical Association. My question is whether these organizations enforce the rules that they themselves have adopted as best practices. I suggest to you that they do not. If they do not regulate themselves in such matters, then it may become necessary to look elsewhere and to have third-party enforcement. The same difficulty has been encountered with the pharmaceutical industry that cannot help but be aware that its products are being acquired and used for purposes that are not therapeutic, but that will do nothing to regulate the sale or distribution of the products for such purposes. Profit seems to be a greater good than ethics. I call on the pharmaceutical industry to help us promote the ethical values of sport. I hope someone with more knowledge of the subject matter will also address the issue of the use of athletes within the research community, to develop performance-enhancing substances and methods that are so clearly antithetical to the spirit of sport. Where will this lead when genetic manipulation becomes a fact of life? This is far more serious than use of anabolic steroids or erythropoietin (epo). How will society deal with the prospect of "breeding" basketball teams in the future? What ethical guidelines will be placed on research and implementation of genetic projects? How will they be enforced? Are there useful distinctions to be made between genetic design (to breed gymnasts) and genetic modification? Or between genetic treatment and genetic enhancement? Can sports leaders be helpful to the scientific community as policies are developed in this field? There are some questions that only experts in this field can answer. You are in the forefront of the new science. You have a better idea than do we, in the sport community, of what lies ahead. You know the risks. You know the policy issues that are involved for society. You know how drugs, such as epo, that have been developed for genuine therapeutic purposes have also been used for mere performance enhancement. You know what issues society as a whole must face. We are prepared to share what we believe are the humanistic issues of sport, as we understand them, but we confess to having no idea where your science will take humanity and how, on the fringes of what may be genuine societal advancement, these advances may be perverted by those who would dehumanize sport. I hope we can emerge from this conference with some guidelines that will be useful in developing a social policy that will distinguish between genuine advancement of the human condition and practices that may lead to the perversion of that condition. All of us look forward to the benefit of your wisdom, experience, and guidance over the next couple of days.

Most of the overall program for the meeting was organized by Theodore Friedmann, Professor of Pediatrics at the University of California San Diego School of Medicine and member of the WADA Health, Medicine, and Research Committee (HMR); Arne Ljungqvist, member of the International Olympic Committee, Chairman of the IOC Medical Commission and Chairman of the WADA HMR Committee; and Gary Wadler, also a member of the WADA HMR Committee and a professor on Medicine at the NYU School of Medicine. Additional planning came from Jan Witkowski, director of the Banbury Center at the Cold Spring Harbor Laboratory on Long Island, New York.

The opening session at the meeting was devoted to summaries of the nature of the doping threat to sport (Arne Ljungqvist, Sweden; Gary Wadler, USA), an athlete's perspective on the extent and depth of the problem (Johan Olaf Koss, Norway), a summary of the present methods for testing and monitoring (H. Klaus Mueller, Germany), and a review of the principles and concepts underlying gene therapy as an impending new approach to doping (Theodore Friedmann, USA). A number of potential approaches to modification of muscle function were described (Bengt Saltin, Denmark; Carl Sandburg, Sweden; Geoffrey Goldsping, United Kingdom; H. Lee Sweeney, U.S.A), and thoughts about pain control (Joseph Glorioso, USA) and joint and orthopedic function (Christopher Evans, USA) were discussed. Other potential models involving energy utilization (Douglas Wallace, USA) and erythropoietin function (Barry Byrne, USA) were presented. Finally, ethical, legal and regulatory issues were presented (Angela Schneider, Canada; Bartha-Marie Knoppers, Canada; Eric Juengst, USA; Odile Haguenauer, France; Richard Young, WADA).

These discussions were particularly valuable because they provided the very first opportunity for the athletic community to hear first-hand from some of the leading scientists in the gene therapy field of the startling prospect for gene-based modification of traits that could affect athletic performance. They also allowed the scientists for the first time to hear first hand from the athletic community of the concerns regarding potential "misuse" of gene transfer technology not to attack disease but rather to modify human traits for the enhancement goal of improving human athletic performance. What emerged from this important meeting was the beginnings of a common language and a common understanding of the nature of the problem. Furthermore, the participants felt strongly that the impetus provided by this first conversation between modern molecular science and those concerned about the abuse of technology for the purpose of sport doping should be carried forward in a broad education program for the scientific and sports communities as well as for the general public through a series of continuing discussion, meetings, and symposia.

To open such a continuing effort, the workshop issued a statement following the meeting, which underscored the importance of the impending problem of gene-based doping and made several recommendations to assure that appropriate scientific and administrative measures are taken to avert the problems. These recommendations included a markedly increased and targeted research effort by the international sports community through WADA and the national anti-doping agencies such as USADA to develop sophisticated molecular and biochemical methods for detecting gene transfer approaches to sports enhancement. The workshop also reiterated the need for an increased program of symposia, publications, and follow-up meetings to educate both the athletic community, as well as the molecular genetics and gene therapy communities to the existence and importance of the problem.

B. Summary of the Banbury Conference conclusions presented to the media following the conference

The conference participants agreed on a series of conclusions, some general and other specific to sports, that were presented in a public press conference after the meeting. The workshop participants concluded the following.

1. General principles

A. *Gene transfer technology, which is still at the investigational stage, is nevertheless already beginning to demonstrate clinical efficacy.*

B. *While genetic technologies hold immense therapeutic promise, there is potential for their misuse, including attempts at the enhancement of athletic performance.*

C. *The collective efforts of scientists, ethicists, athletes, sports authorities, medical practitioners, professional societies, pharmaceutical and biotech industries, and public authorities (including governments) will be required to avert such misuse.*

D. *The compliance with established international standards pertaining to genetic experimentation involving human subjects, such as the Helsinki, Geneva, and Lnuyama Declarations that prevent unethical research is essential. The application of genetic transfer technologies should be consistent with established standards of professional behavior.*

E. *The pace of research in the field of genetic transfer technology is such that governmental and other regulatory agencies must work with a continued sense of urgency to establish a social and policy framework to guide this research and its applications and sanction breaches of the framework.*

F. *Broad public discussion and the development of social and policy frameworks must surround the distinction between genetic therapy and genetic enhancement. The time for the social framework to be established is before abuses occur, not after-the-fact.*

2. Sports specific

A. *Athletes, in common with other people in society, are entitled to the benefits of genuine therapeutic applications to treat injuries and other medical conditions.*

B. *There are evident risks that genetic transfer technologies might be used in a manner that would be contrary to the spirit of sports or potentially dangerous to the health of athletes. Akin to doping in the present generation, genetic transfer technology that is nontherapeutic and merely performance enhancing should be prohibited.*

C. *The definition of doping used by WADA, the IOC, international sports federations (IFs), and national authorities should be expanded to include the unapproved use of genetic transfer technologies.*

D. *One of the benefits of genetic technology is its potential use in the detection of prohibited substances and methods.*

E. *The scientific community has recognized the need for the continued development and refinement of methods that will permit the detection of the misuse of genetic transfer technologies in sports. The conference noted there are a number of approaches that currently exist, or are in development, that will permit such detection.*

F. *The present focus of WADA's research grants toward the study of the detection methods for the misuse of oxygen-carrying agents and growth factors should be extended to include the detection of genetic transfer technologies and their effects.*

G. *The World Anti-Doping Code, which is planned for implementation by 2004, should include language prohibiting the use of genetic transfer technologies to enhance athletic performance.*

H. *WADA calls upon its government members, in particular, to expedite the development of a global social framework for the application of genetic transfer technologies that address the potential misuse of these technologies in sports and a publicly stated deadline for the adoption of that framework.*

I. *WADA calls upon governments to consider the following recommendations for inclusion in the regulatory framework pertaining to genetic transfer technologies and related research:*

1. *Address breaches of the social framework within the criminal or penal realm;*
2. *Extend corporate liability to directors, officers, and senior employees;*
3. *Extend civil and criminal limitation periods in respect of breaches of the regulatory framework;*

4. *Require detailed record-keeping in respect of all applications of gene transfer technologies with independent audit requirements; and*
5. *Expand standards of medical and professional behavior to prohibit the improper use of genetic transfer technologies and that such rules be actively enforced;*

J. *WADA calls upon governments and the sports movement to establish and fund educational and ethics programs designed to prevent the possible misuses of genetic transfer technologies in sport. WADA is willing to coordinate the design and dissemination of such programs.*

K. *WADA and the scientific community will establish a mechanism for continuing dialogue and consultation around the subject of genetic transfer technologies.*

II. BEYOND BANBURY

A number of the recommendations stemming from the Banbury meeting have been implemented. Perhaps the most impressive has been the development of rigorous research programs by WADA and USADA aimed at the identification of molecular markers for exposure to doping substances. These programs have not been designed to replace the current program that emphasizes the use of the kinds of tests typical of most current modern testing and monitoring procedures used in sport doping. These classical and well-established methods revolve largely around developing specific analytical tests for the offending drug and biochemical tests of the physiological effects of drugs. For instance, monitoring for steroid use often requires a new biochemical test for each new steroid. Detection of erythropoietin doping involves not only the demonstration of the foreign erythropoietin itself in urine but also detection of the physiological effects of the hormone—the increased number of red blood cells in the circulation. Instead, and largely as a result of the impetus provided by the Banbury workshop, the augmented research programs at WADA and USADA now include studies designed to detect universal or, at least, very common, changes in the expression of many genes in response to exposure of an athlete to a drug—changes that would constitute a molecular "signature" for exposure to a class of drugs. As an example, it is presumed, with good justification from many preliminary studies, that anabolic steroids of the kind banned in sport bring about their effect in tissues by changing the ways in which many, many genes are expressed—genes that control growth of muscle and other tissues, genes that regulate the way in which tissues use energy, genes that determine how neuromuscular signals are converted into muscle contraction, and so on. As a result of the explosive growth of molecular genetics in the past several decades, spurred by the technological needs of the human genome project, extremely powerful methods are now available to look simultaneously at the extent to which all 20,000–30,000 human genes are expressed and how their expression is altered by powerful hormones and other drugs that affect muscle growth and function. This startling technique, called "microarray analysis", uses a small "chip" that contains all 20,000–30,000 human genes on a glass or silicon chip less than one inch square, which can then be reacted with chemicals that emit light signals in proportion to the extent to which the gene is expressed. In just several days, one can determine how the entire collection of human genes responds to exposure to any agent—a drug, an environmental toxin, etc. This is precisely the approach that is also being taken so effectively to determine the differences, for instance, between normal and cancer cells. In addition, it has become possible through an approach called "proteomic analysis" to examine simultaneously all of the thousands of proteins that are expressed by genes and that are found in blood, urine, other body fluids and in all cells. If one imagines how these and other immensely powerful new detection methods can be useful in sport doping, envision the great difficulty posed by new, designer" steroids that are created by rogue chemists precisely to avoid detection because they are "invisible" to existing chemical tests. It is safe to presume that most or even all anabolic steroids will all use common pathways to achieve their effect, and those effects will be reflected in changes in gene expression and in the distribution of proteins (the "proteome") of affected tissues. Toward that end, at least some of the new WADA and USADA research programs are built on this premise and it's safe to say that very promising results are beginning to appear that lend support to the underlying concept.

If this approach proves effective, it would go a long way toward the goal of developing relatively noninvasive testing and monitoring procedures for drug testing and monitoring in sport doping. It is of great interest to the gene therapy community to keep track of genes that are introduced into subjects and patients, to determine how well, how long and at what levels foreign genes are expressed. Acquisition of this kind of information often requires relatively invasive procedures such as tissue biopsies, and of course such a procedure would not be feasible in the context of sport and monitoring for sport doping. If the introduction of foreign genetic material expressing, for instance, a growth-promoting function such as growth hormone, into a specific muscle were to be shown to produce genetic or proteomic changes, for instance, in circulating blood cells or in blood itself, then the difficulties stemming from invasiveness of tissue biopsies and from the difficulty in knowing where a biopsy should be done would be avoided. One could in principle detect a doping event even if one had no idea where the gene was introduced and, to some extent, the exact structure of the drug. We have confidence that such screening tests are feasible and will emerge from the WADA and USADA and other similar research programs.

7 A Scientific Perspective

I. GENE THERAPY IS A REALITY

It seems very clear that the field of human gene therapy, despite a difficult gestation, has now been born and is a reality. It is a reality not only as a new and quite epochal concept in Medicine but also, even more importantly, it is a reality in successful clinical translation from the bench to the bedside. There was never any doubt that sooner or later, it was going to be possible to transfer genes or pieces of genetic information into defective human cells and correct or prevent many kinds of human diseases, from the very rare disease "orphan" diseases that, while causing immense pain and suffering in small numbers of patients, have little broad social impact to the very common diseases—cancer, heart and other cardiovascular diseases, degenerative neurological diseases,

Advances in Genetics, Vol. 51 0065-2660/06 $35.00

DOI: 10.1016/S0065-2660(06)51007-7

etc., that are the major health burdens and killers in our society. So sure were most of us that we as a community became too optimistic about how simple it was going to be to translate the concept of gene transfer from early laboratory findings to human beings. As a community, and mostly for the best of intentions, we downplayed the scientific, policy, and ethical difficulties; we exaggerated the benefits and the ease of achieving them. We cried "wolf" too often and too loudly. As a community, we did so largely because of our fervent belief that this direction was crucial for the development of treatments for otherwise intractable diseases. Historically, an emphasis on achievability and quick results has often been necessary to secure funding from the traditional funding agencies, commercial partners, and charitable foundations. In the case of the emerging early field of gene therapy, there was a high level of interest in the private sector—the investment and venture capital communities and the biotechnology and pharmaceutical industries. Through these avenues, large amounts of funding became available, at least temporarily, but were often tied to clinical application and the promise of achievable, rapid, and even lucrative results. During the early formative days of the field, many symposia and scientific meetings, communications with patients' groups and with the general public, and interactions with the biotechnology and pharmaceutical worlds emphasized and overemphasized the seductively attractive early findings in laboratory models and even in animal studies. We became prepared only for the expected good results when the methods were applied to human patients.

That is why the field was so startled by and unprepared for the series of scientific setbacks that started to come in the mid-1990s. In 1995, two advisory committees reported to the director of the US National Institutes of Health that the field of gene therapy was full of promises but was also beset by too many unfulfilled and exaggerated promises. That is obviously also the basis for the severe blow to the confidence that many well-meaning and concerned people—scientists in and out of the field, public policy officials, and the media—had all placed in the field that came after the death of Jesse Gelsinger, from the cases of leukemia in the X-SCID study, and from a number of additional surprising negative clinical results not described in detail here. That is also why the field of gene therapy continues to be viewed by many in the general media and even in parts of the scientific community with a high level of skepticism and disbelief. It has become too easy to dismiss the real progress as just more unfulfillable promises. For instance, it is well known that many clinical studies carried out by pharmaceutical companies are abandoned at all stages—early and even very late—of their development because results indicate that the approach does not work or harmful. Most such studies are discontinued quietly and without great fanfare or public notice, even after expenditure by the companies of the large amounts of money required to shepherd a new treatment through the long and rigorous development process and through the regulatory hurdles. However, because of the aura of disbelief hanging over the field of gene therapy, a decision by a commercial sponsor to discontinue an important study of gene therapy for a form of hemophilia because of safety concerns was met with notices in the media of yet another *failure* for the field of gene therapy rather than with an acknowledgement that such a shutdown represents the normal process of revaluation and reevaluation of clinical studies and discontinuation of those in which serious safety issues arise. It was in fact far less "another failure" than it was an appropriate and positive move by the company to protect patients.

But the field had shot itself in the foot a number of times, and it is no great surprise that it is being held to a different standard than the one it held itself to just a few years ago. Results were promised too quickly, too easily. When one considers the time required to move from initial concepts of gene therapy in the early 1970s to convincing clinical therapy in 2002—three decades—one might conclude that represents a very long time. However, that kind of long development time to bring a new therapy from concept to wide-scale delivery is typical for many difficult and complex new concepts and therapies in medicine; it can take several decades for bold new steps to become established and effectively applied to patients. For instance, the initial concepts and techniques of chemotherapy for several of today's most treatable cancers—some kinds of childhood leukemia, forms of Hodgkin's disease, etc.—required two to three decades to mature and improve cure rates from several percent to the level possible now—more than 80–90% cures for these previously lethal diseases.

Gene therapy is an epochal but difficult new form of therapy, and it is no surprise whatsoever that it has taken more than a decade of clinical studies to prove effective in the experimental setting. It is now reasonable to move away from an obsession with the difficulties, the setbacks, and the failures and finally recognize the immense achievement of the field—a new form of medicine has been born. It will surely move with increasing speed to wide-scale application. Like all newborns, there are many things it cannot yet do. They will all come in time.

II. THE LINE BETWEEN THERAPY AND ENHANCEMENT IS INDISTINCT: WHAT DESERVES "TREATMENT"?

One of the areas in which troublesome scientific, policy, and ethical issues related to gene transfer applications in humans is the indistinct border between therapy and enhancement. What trait is being modified? Is it a disease or the cause of a severely compromised quality of life? Is the modification justified? There are certainly human conditions that most of us—with the exception of adherents to some religious principles—would agree are "diseases" and require intervention. Not many could look upon children afflicted with severe combined immunodeficiency disease (SCID), unable to live outside of their protective plastic bubbles and without aggressive treatment of the life-threatening infections during their short lives, and not conclude that this is disease that requires us to act. Not many of us would be able to see a blind infant and not wish to reverse that damage and "cure" that child. Those conditions, most of us would agree, are diseases and must be railed against.

There are other physical traits that are not so clearly "diseases" but that many in our society nonetheless might wish to modify for the purpose of improving "quality of life" or even for cosmetic reasons. Most human characteristics span a broad range from obviously "defective" to more-or-less "normal," and most of our modern societies have already accepted the notion that the edges of this normal distribution that border on the "abnormal" or "defective" are targets for modification. For instance, there is a very wide variation in the height of human beings—

some are very tall, some are very short. In some cases, such extremes result from demonstrable defects in growth factors, such as growth hormone, leading to pathological shortness. These children are often treated with growth hormone to add valuable inches to their eventual height. In many other instances, people are short in the absence of any demonstrable abnormalities in these or other growth factors. They are short because their parents are short—through whatever mechanisms that may reflect. Such a trait is just as clearly inherited as is an inborn genetic deficiency of growth hormone, but the resulting short stature might not be seen by most of us to be abnormal or require therapy. Programs for treating short children formally are base on a demonstration that there is a deficiency of growth hormone in such children. However, it is well known that many children without growth hormone deficiency are treated with growth hormone because the added several inches of height is seen as a social benefit to some children—not for the purpose of making them potential basketball starts but to spare them the taunts and teasing that some such short children face from their peers.

The loss of muscle tone and strength during aging is "normal"—it happens to all sooner or later. But is it a condition that requires "treatment"? At least in animal-model studies, muscle tone and strength certainly seem to benefit from intervention—injections of growth factors such as IGF-1 or of the gene that produces IGF-1, or possibly other factors. It is very likely to be just as true of aging humans and treatment can be aimed at quality of life rather than at a specific genetic disorder, with potentially helpful results to many aging but otherwise normal people.

Humans also often show a loss of memory and intellectual acuity during "normal" aging. In the extreme forms of these problems found Alzheimer's disease, we have taken the extreme nature of the changes to represent pathology and real disease, and a great deal of research is being aimed at treatment for this disorder. And yet many will be tempted to apply drugs or even gene-based manipulations developed for Alzheimer's treatment for the far less severe but nonetheless troublesome dwindling of those functions in all of us as we age "normally."

Even complex social behaviors are now slowly coming to be understood as reflections of mixed genetic and environmental influences.

For instance, mice having a deficiency in the gene known as *MECP2* have been developed as a model for the human autism-like disorder called Rett's syndrome. Mice with this genetic deficiency show deficiencies in a variety of behavioral traits important to mice-nesting behavior and reduced decisiveness in interacting with other mice. Similarly, mice with defects in another gene (i.e., *Dvl1*) also demonstrated complex changes in socialization behavior and deficits in the ability to social hierarchies and dominance. A study in mice demonstrated that defects in the gene making the protein *stathmin* seemed to make timid mice much more daring and less intimidated by novel situations, possibly by interfering with nerve cell connections in the part of the brain called the amygdala, a center thought to be important in fear responses. And so it seems likely that, as this kind of research gathers speed, more genes will be found with powerful effects on complex social behaviors and thereby make those genes susceptible to manipulation and possibly make these and many other normal human behaviors available for drug- and possibly genetic intervention.

We already have numerous examples of widely accepted "*treatments*" aimed at enhancement of normal human traits for improving quality of life—we are normal but not normal enough. We have medicine cabinets full of drugs designed to improve our moods and memory, enhance our sexual performance, and modify aspects of our personality and temperament. If drugs are acceptable for these purposes, why not genetic approaches to exactly the same conditions? It might be difficult to develop convincing and logically consistent arguments against such applications of gene transfer. In fact a number of eminent scientists, including most notably James Watson, the discoverer with Francis Crick of the structure of DNA, have argued that it is unethical to have these tools available to us and not to use them to improve humans through genetic modifications. These issues have not been faced squarely by our society and, as in many instances, we may have to wait until scientific or public policy disasters occur before we can come to any social consensus on how to face these potential problems. As described in an earlier chapter, the extensive infrastructures in place to oversee and regulate most but not necessarily all gene transfer applications in human subjects and patients have for now elected not to accept enhancement studies.

But surely, we will not have the option of delaying that discussion for very long; enhancement of gene transfer applications will surely come as the technology improves and no longer poses the safety issues that still hover over all genetic modification studies at the moment.

III. REPAIR OF INJURY IS A SPECIAL FORM OF ENHANCEMENT IN SPORTS

The effective management of injury and the repair of damaged tissues are central to athletic training and competition. A great deal of research is being carried out to develop methods to speed and improve tissue repair by introducing genes into injured muscles, tendons, joints, etc. that stimulate growth and repair. In fact, studies in animal models have shown that the gene transfer into such damaged tissues can speed repair and regrowth of some injured tissues—muscle, tendon, and cartilage. Certainly, if these kinds of treatments are found to be effective in human injuries, athletes should not be deprived of the therapeutic benefits of gene-base repair of legitimate athletic injuries. However, one can also imagine a temptation to perform preventive treatment of vulnerable or repeatedly injured tissues not to treat but to prevent future injury. These kinds of manipulations of course need not be restricted to human athletes but could readily be imagined to be equally possible, for example, in the highly vulnerable limbs and joints of race horses.

IV. MISUSE OF GENETIC SCIENCE: THE REEMERGENCE OF EUGENICS

Through the genetics revolution that we have been witnessing during the past several decades, we have come to realize not only that most human disease is determined at least partly by genetic components but also that most normal human traits are similarly affected to a greater or lesser degree by genetic influences. These include our most complex and mysterious traits—our personalities and cognitive and intellectual capacities. Of course, most of the genes involved in these functions have not yet been identified but we know that they are there. We do not yet know what they are or what they do, but in time we will find them and increasingly understand them.

The tools being developed to achieve safe and efficient genetic attack on disease-related genes are all too uncomfortably applicable to the manipulation of genes associated not with disease but with normal human traits. The idea that these genetic properties should be subject to manipulation of one sort or another is not new, and a very powerful movement arose especially in the UK and the United States in the late 1800s and early 1900s based on the idea that many traits, such as poverty, slovenliness or laziness, prostitution, many kinds of mental deficiency, and many others considered socially undesirable were inherited and that they posed a threat to the genetic health of the human species. Especially in the United States, this concept based on poor science was too effectively translated into discriminatory and evil social policy, such as restrictive immigration policies and even forced sterilization programs for mental retardation and some kinds of neurological disease. So powerful was the idea that undesirable human traits posed dangers to the American society that these terrible social policies were supported by the United States Supreme Court in a decision handed down in the 1927 case *Buck* v. *Bell*. The court affirmed previous lower court decisions in favor of laws in the State of Virginia that allowed forced sterilizations of a woman because several members of her family were thought—not proven—to be feebleminded. The famous opinion of Chief Justice Oliver Wendell Holmes declared that, in order to "prevent the nation from being swamped with incompetence" sterilization was warranted .... "Three generations of imbeciles are enough," declared Justice Holmes. It has been very well documented that the harmful and misguided social and political responses to eugenic concepts by some geneticists, social scientists, politicians, and public policy figures in many countries but most powerfully in the United States, helped to shape the eventual horrors of Nazi Germany.

Certainly the scientific and ethics climates are far different from what they were a century ago. Science is much more solid and rigorous now and the understanding of the relevant genes and their roles in normal human functions, if not very precise yet, will certainly be much better than the pseudo-scientific ideas of the early 1900s that led to *Buck v. Bell* and to the Nazi programs. Given the much better science available now and the greater awareness of ethical requirements for appropriate manipulation of human biology that we have compared with past generations, we could never repeat the genetic abuses and missteps of the last century—could we?

To the extent that enhancement in sports might easily represent one of the earliest opening scenarios in the broad problem of gene-based enhancement, what occurs in the world of sports is likely to define the ways in which our society approaches and solves the problems arising from the deliberate genetic modification of human traits in general.

8

Some Ethical Aspects of "Harm" in Sport

Some of the arguments against all forms of doping, including genetic doping in sport revolve around the various forms of harm that would ensue, as introduced in Chapter 1. At least four types of harm or potential harm can generally be envisioned including: (1) harm to users, (2) harm to other athletes, (3) harm to society, and (4) harm to the sports community. The most obvious kind of harm would result from the known harmful effects of many of the drugs classically used in sport doping in general and from the scientific and medical unknowns that still permeate the field of gene transfer in human subjects and patients in the gene therapy setting. We examine here some of the ethical principles that underlie the response to these kinds of real and potential harm.

I. HARM AND HEALTH: THE INDIVIDUAL ATHLETE

One of the common ethical arguments justifying a ban on doping in sport is the potential for harm to the athlete, that the athlete requires protection, that such protection could be provided by banning the offending

Advances in Genetics, Vol. 51

0065-2660/06 $35.00
DOI: 10.1016/S0065-2660(06)51008-9

substance or procedure. The assertion that the use of gene transfer technology for enhancement will harm the user is not unreasonable. We have seen evidence in earlier discussions that the technology of gene transfer even in the case of therapy for dire disease is far from perfect and that a number of studies have resulted in severe damage—even death—to research subjects and patients. We all accept many risks in clinical research and human experimentation in the name of easing the pain and suffering of disease, with the perspective of improved well-being or of a cure, but are generally loathe to accept the same adverse events if the subject were a healthy young athlete. We feel the need to protect research subjects from harm.

It seems reasonable to suppose that the likelihood of harm likely to befall a healthy athlete or other healthy subject, taking doping substances or taking part in a gene transfer approach to doping without medical indication, is greater than the likelihood of harm that would befall someone suffering from an illness for which the same agent represents a therapy. At least the ratio of harm to benefit would be far greater in normal young subjects than in ill patients. For instance, treatment of males suffering from hypogonadism or age-related testosterone deficiency with testosterone replacement is medical therapy with a real risk/benefit ratio. Treatment of healthy young athletes with the same substance probably carries a far greater risk/benefit ratio. Is it justifiable to protect an athlete from such risks?

The desire to act on behalf of others or to protect others from the consequences of their own actions is known as paternalism. Obviously, in the case where potentially injurious actions are being undertaken by minors or people who are otherwise medically and medico-legally incompetent because of illness, age, or injury, paternalism is acceptable, legal, and ethical. Virtually everyone can accept the need for drug and genetic regulation and even banning of doping by minors. But most athletes are not incompetent—most are adults who are fully competent from all ethical and medico-ethical perspectives.

Almost all of us, except the most extreme libertarians among us who consider any form of governmental intervention in personal lives of people to be unacceptable, accept the notion that there are ethically acceptable forms of paternalism, even toward competent people like health young athletes. We accept forms of paternalistic governmental behavior—we prohibit driving without seatbelts and drinking and driving.

On the other hand, under some circumstances paternalism toward competent people able to act on their own behalf can be antithetical to one of the primary precepts of ethical behavior—protection of the primacy of an individual's autonomy. In fact, much of the thrust of modern North American medical ethics has been directed precisely against medical paternalism. We treasure and protect the concept that most adult individuals should be considered to be competent and entitled to make their own decisions.

Although most elite athletes are competent adult athletes, some argue that banning doping by them could be considered a form of acceptable paternalism—an attempt to protect the health or well-being of a competent adult athlete or to prevent some of the other forms of harm that result from doping. On the other hand, paternalistic interventions in the lives of competent adult athletes can be seen by others to be unwarranted—that others know better than the athletes themselves how to achieve a more general good that could deny for them the very attributes that we so highly value—self-reliance, personal achievement, and autonomy. But of course the flaw in that argument is that sports ethics is not the same as medical ethics. Drug bans are meant not only to protect athletes from harm but also to preserve the concept and ethos of sport—the concept that rules of athletic competition are established for those who would wish to participate and that those who choose to take part do so as volunteers and agree to bans and monitoring programs as the price one pays for the privilege—not the right—to participate in competitive sport.

We can therefore conclude that the argument in favor of drug bans based largely on the need to protect athletes from harm is an important but an incomplete justification. However, in the case of a technology as immature as gene transfer approaches to sport doping, we find a particularly convincing argument of athlete protection. The state of the technology underlying this approach to doping is extremely immature and full of recognized and even more unrecognized dangers. Any suggestions that we know enough at the moment to guarantee effectiveness of such procedures as well as safety would be dishonest and deceptive and missing the essentials of acceptable experimental work with human subjects. Autonomous choices are autonomous only if they are enlightened and that a gene-based doping application under the current conditions of inadequate scientific knowledge and probable secrecy and stealth that would almost certainly be involved, a decision to protect athletes from such inappropriate genetic applications would certainly be an example of acceptable paternalism. Once the tools and methods of gene therapy will have become much more fully established and proven to be safe, the issue of personal safety by itself will become a much less convincing justification for a ban on genetic doping.

II. HARM TO OTHER ATHLETES

The second form of harm from drug- or gene-based doping is not the harm that gene transfer technology could cause to users, but rather the harm that could befall other athletes. The "others" in this argument are usually deemed to be other "clean" (undoped) athletes and the harm comes through "coercion." If a successful athlete is known to be doping, then others who wish to compete and succeed at that level might feel compelled to resort to the same kind of illicit methods to have a realistic prospect for success. Thus, doping is a harm because it is coercive and puts the well-being and life goals of others in jeopardy. Competitors require protection from the risks of doping and from the damage caused to them and to their careers. In principle such protection might be provided by a ban on the substance or the technique in question.‘

III. HARM TO SOCIETY

Any form of doping, whether it be—drug- or gene-based, also may harm another group of people—the general public and, in particular, children. People admire prominent successful athletes and view them as role models. We often try to emulate the nobler activities and qualities of great athletes—their struggles against great odds, their tenacity, and their dedication to a goal—and extrapolate their achievements to many non-sports domains of our society. Whether this is a fair or an appropriate burden to put onto athletes rather than other public figures is irrelevant—it is a fact. The trust that we place onto athletes seems to be a societal "good" and anything that brings that trust into question or into disrepute must be a form of harm. Sport is one of the very first areas in which young people hope to gain excellence, the excellence of their heroes and heroines. From a societal perspective, if the hero or heroine is tainted and becomes morally suspect, the very young may have difficulty distinguishing between the athletic triumphs of their heroes or heroines and the moral or ethical blots on them. The enobling influence of the hero becomes a crushing negative influence of a fallen idol. Interestingly, the downfall of a hero might not be intrinsic to doping itself but rather could merely be due to the revelation of doping that results from breaking the rules of the sport. If the rule does not exist, the fall from grace would possibly also not be as severe. For instance, if it was publicly revealed that a prima ballerina used painkillers or stimulants to achieve an excellent performance of "Swan Lake," we would not hear the same outcry because rules do not exist against the use of such drugs in artistic achievement. Our society's expectations on high-level athletes therefore seem to be different from that placed many of our other public figures, and it should therefore not be a surprise that our societal responses to their "misdeeds" and missteps will be different.

9 The List

Presented below is the WADA list of banned substances and methods that takes effect at the beginning of 2006. It should be remembered that this is a constantly evolving document that responds quickly as new scientific and medical information becomes available. WADA maintains an up-dated version of the list in both English and French on its web site (http://www.wada-ama.org/en/) and all concerned parties—athletes, trainers, athletic handlers, medical practitioners, and scientists should refer to that source of latest information.

Advances in Genetics, Vol. 51 0065-2660/06 $35.00

DOI: 10.1016/S0065-2660(06)51009-0

The World Anti-Doping Code

THE 2006 PROHIBITED LIST
INTERNATIONAL STANDARD

This List shall come into effect on 1 January 2006.

THE 2006 PROHIBITED LIST
WORLD ANTI-DOPING CODE

The use of any drug should be limited to medically justified indications

SUBSTANCES AND METHODS PROHIBITED AT ALL TIMES (IN- AND OUT-OF-COMPETITION)

PROHIBITED SUBSTANCES

S1. ANABOLIC AGENTS

Anabolic agents are prohibited.

1. Anabolic Androgenic Steroids (AAS)

a. Exogenous[1] **anabolic androgenic steroids** (AAS) include the following:

1-Androstendiol (5α-androst-1-ene-3β,17β-diol); **1-androstendione** (5α-androst-1-ene-3,17-dione); **bolandiol** (19-norandrostenediol);

[1]For this section "exogenous" refers to a substance, which is not ordinarily capable of being produced by the body naturally.

bolasterone; boldenone; boldione (androsta-1,4-diene-3,17-dione); **calusterone; clostebol; danazol** (17α-ethynyl-17β-hydroxyandrost-4-eno[2,3-d] isoxazole); **dehydrochlormethyltestosterone** (4-chloro-17β-hydroxy-17α-methylandrosta-1,4-dien-3-one); **desoxymethyltestosterone** (17α-methyl-5α-androst-2-en-17ß-ol); **drostanolone; ethylestrenol** (19-nor-17α-pregn-4-en-17-ol); **fluoxymesterone**; **formebolone**; **furazabol** (17β-hydroxy-17α-methyl-5α-androstano[2,3-c]-furazan); **gestrinone; 4-hydroxytestosterone** (4,17β-dihydroxyandrost-4-en-3-one); **mestanolone; mesterolone; metenolone; methandienone** (17σ-hydroxy-17α-methylandrosta-1,4-dien-3-one); **methandriol; methasterone** (2α,17α-dimethyl-5α-androstane-3-one-17β-ol); **methyldienolone** (17β-hydroxy-17α-methylestra-4,9-dien-3-one); **methyl-1-testosterone** (17β-hydroxy-17α-methyl-5α-androst-1-en-3-one); **methylnortestosterone** (17β-hydroxy-17α-methylestr-4-en-3-one); **methyltrienolone** (17β-hydroxy-17α-methylestra-4,9,11-trien-3-one); **methyltestosterone; mibolerone; nandrolone; 19-norandrostenedione** (estr-4-ene-3,17-dione); **norboletone; norclostebol; norethandrolone; oxabolone; oxandrolone; oxymesterone; oxymetholone; prostanozol** ([3,2-c]pyrazole-5a-etioallocholane-17β-tetrahydropyranol); **quinbolone; stanozolol; stenbolone; 1-testosterone** (17β-hydroxy-5α-androst-1-en-3-one); **tetrahydrogestrinone** (18a-homo-pregna-4,9,11-trien-17β-ol-3-one); **trenbolone**, and other substances with a similar chemical structure or similar biological effect(s).

b. Endogenous[2] AAS includes the following:

Androstenediol (androst-5-ene-3β,17β-diol); **androstenedione** (androst-4-ene-3,17-dione); **dihydrotestosterone** (17β-hydroxy-5α-androstan-3-one); **prasterone** (dehydroepiandrosterone, DHEA); and **testosterone**. It also includes the following metabolites and isomers.

5α-Androstane-3α,17α-diol; 5α-androstane-3α,17β-diol; 5α-androstane-3β,17α-diol; 5α-androstane-3β,17β-diol; androst-4-ene-3α,17α-diol; androst-4-ene-?α,17β-diol; androst-4-ene-3β,17α-diol;

[2]For this section "exogenous" refers to a substance, which is not ordinarily capable of being produced by the body naturally.

androst-5-ene-3α,17α-diol; androst-5-ene-3α,17β-diol; androst-5-ene-3β,17α-diol; 4-androstenediol (androst-4-ene-3β,17β-diol); 5-androstenedione (androst-5-ene-3,17-dione); epi-dihydrotestosterone; 3α-hydroxy-5α-androstan-17-one; 3β-hydroxy-5α-androstan-17-one; 19-norandrosterone; and 19-noretiocholanolone.

Where an anabolic androgenic steroid is capable of being produced endogenously, a *sample* will be deemed to contain such *prohibited substance* where the concentration of such *prohibited substance* or its metabolites or markers and/or any other relevant ratio(s) in the *athlete*'s *sample* so deviates from the range of values normally found in humans that it is unlikely to be consistent with normal endogenous production. A *sample* shall not be deemed to contain a *prohibited substance* in any such case where an *athlete* proves that the concentration of the *prohibited substance* or its metabolites or markers and/or the relevant ratio(s) in the *athlet's sample* is attributable to a physiological or pathological condition.

In all cases, and at any concentration, the *athlete*'s sample will be deemed to contain a *prohibited substance* and the laboratory will report an *adverse analytical finding* if, based on any reliable analytical method (e.g., IRMS), the laboratory can show that the *prohibited substance* is of exogenous origin. In such case, no further investigation is necessary.

If a value in the range of levels normally found in humans is reported and the reliable analytical method (e.g., IRMS) has not determined the exogenous origin of the substance, but if there are serious indications, such as a comparison to reference steroid profiles, of a possible *use* of a *prohibited substance*, further investigation shall be conducted by the relevant *anti-doping organization* by reviewing the results of any previous test(s) or by conducting subsequent test(s), in order to determine whether the result is due to a physiological or pathological condition, or has occurred as a consequence of the exogenous origin of a *prohibited substance*.

When a laboratory has reported a T/E ratio greater than four (4) to one (1) and any reliable analytical method (e.g., IRMS) applied has not determined the exogenous origin of the substance, further investigation may be conducted by a review of previous tests or by conducting subsequent test(s), in order to determine whether the result is due to a physiological or pathological condition, or has occurred as a consequence of the exogenous origin of a *prohibited substance*. If a laboratory reports, using an additional reliable analytical method (e.g., IRMS), that the *prohibited substance* is of exogenous origin, no further investigation is necessary and the *sample* will be deemed to contain such *prohibited substance*.

When an additional reliable analytical method (e.g., IRMS) has not been applied and a minimum of three previous test results are not available, the relevant *anti-doping organization* shall test the *athlete* with no advance notice at least three times within a 3-month period. If the longitudinal profile of the *athlete* that is subject to the subsequent tests is not physiologically normal, the result shall be reported as an *adverse analytical finding*.

In extremely rare individual cases, boldenone of endogenous origin can be consistently found at very low nanograms per milliliter (ng/mL) levels in urine. When such a very low concentration of boldenone is reported by a laboratory and any reliable analytical method (e.g., IRMS) applied has not determined the exogenous origin of the substance, further investigation may be conducted by a review of previous tests or by conducting subsequent test(s). When an additional reliable analytical method (e.g., IRMS) has not been applied, a minimum of three no advance notice tests in a period of three months shall be conducted by the relevant *anti-doping organization*. If the longitudinal profile of the *athlete* who is subject to the subsequent tests is not physiologically normal, the result shall be reported as an *adverse analytical finding*.

For 19-norandrosterone, an *adverse analytical finding* reported by a laboratory is considered to be scientific and valid proof of exogenous origin of the *prohibited substance*. In such case, no further investigation is necessary.

Should an *athlete* fail to cooperate in the investigations, the *athlete's sample* shall be deemed to contain a *prohibited substance*.

Other anabolic agents, including but not limited to are: **clenbuterol, tibolone, zeranol, and zilpaterol**.

S2. HORMONES AND RELATED SUBSTANCES

The following substances, including other substances with a similar chemical structure or similar biological effect(s), and their releasing factors, are prohibited:

1. Erythropoietin (EPO)
2. Growth hormone (hGH), insulin-like growth factors (e.g., IGF-1), mechano growth factors (MGFs)
3. Gonadotrophins (LH, hCG), prohibited in males only
4. Insulin
5. Corticotrophins

Unless the *athlete* can demonstrate that the concentration was due to a physiological or pathological condition, a *sample* will be deemed to contain a *prohibited substance* (as listed above) where the concentration of the *prohibited substance* or its metabolites and/or relevant ratios or markers in the *athlete*'s *sample* so exceeds the range of values normally found in humans that it is unlikely to be consistent with normal endogenous production.

If a laboratory reports, using a reliable analytical method, that the *prohibited substance* is of exogenous origin, the *sample* will be deemed to contain a *prohibited substance* and shall be reported as an *adverse analytical finding*.

The presence of other substances with a similar chemical structure or similar biological effect(s), diagnostic marker(s) or releasing factors of a hormone listed above or of any other finding which indicate(s) that the substance detected is of exogenous origin, will be deemed to reflect the use of a *prohibited substance* and shall be reported as an *adverse analytical finding*.

S3. BETA-2 AGONISTS

All beta-2 agonists including their D- and L-isomers are prohibited. As an exception, formoterol, salbutamol, salmeterol, and terbutaline, when administered by inhalation, require an abbreviated therapeutic use exemption.

Despite the granting of any form of therapeutic use exemption, a concentration of salbutamol (free plus glucuronide) greater than 1000 ng/mL will be considered an *adverse analytical finding* unless the athlete proves that the abnormal result was the consequence of the therapeutic use of inhaled salbutamol.

S4. AGENTS WITH ANTIESTROGENIC ACTIVITY

The following classes of antiestrogenic substances are prohibited:

1. Aromatase inhibitors including, but not limited to, anastrozole, letrozole, aminoglutethimide, exemestane, formestane, testolactone.
2. Selective estrogen receptor modulators (SERMs) including, but not limited to, raloxifene, tamoxifen, and toremifene.
3. Other antiestrogenic substances including, but not limited to, clomiphene, cyclofenil, and fulvestrant.

S5. DIURETICS AND OTHER MASKING AGENTS

Masking agents include but are not limited to:

Diuretics,[3] epitestosterone, probenecid, alpha-reductase inhibitors (e.g., finasteride, dutasteride), plasma expanders (e.g., albumin, dextran, and hydroxyethyl starch).

Diuretics include:

Acetazolamide, amiloride, bumetanide, canrenone, chlorthalidone, etacrynic acid, furosemide, indapamide, metolazone, spironolactone, thiazides (e.g., bendroflumethiazide, chlorothiazide, and

[3]A therapeutic use exemption is not valid if an athlete's urine contains a diuretic in association with threshold or subthreshold levels of a prohibited substance(s).

hydrochlorothiazide), triamterene, and other substances with a similar chemical structure or similar biological effect(s) (except for drosperinone, which is not prohibited).

PROHIBITED METHODS

M1. ENHANCEMENT OF OXYGEN TRANSFER

The following are prohibited:

a. Blood doping, including the use of autologous, homologous, or heterologous blood or red blood cell products of any origin.
b. Artificially enhancing the uptake, transport, or delivery of oxygen, including but not limited to perfluorochemicals, efaproxiral (RSR13), and modified hemoglobin products (e.g., hemoglobin-based blood substitutes, microencapsulated hemoglobin products).

M2. CHEMICAL AND PHYSICAL MANIPULATION

a. *Tampering*, or attempting to tamper, in order to alter the integrity and validity of *samples* collected during *doping controls* is prohibited. These include but are not limited to catheterization, urine substitution and/or alteration.
b. Intravenous infusions are prohibited, except as a legitimate acute medical treatment.

M3. GENE DOPING

The nontherapeutic use of cells, genes, genetic elements, or of the modulation of gene expression, having the capacity to enhance athletic performance, is prohibited.

SUBSTANCES AND METHODS PROHIBITED IN COMPETITION

In addition to the categories S1–S5 and M1–M3 defined previously, the following categories are prohibited in competition.

PROHIBITED SUBSTANCES

S6. STIMULANTS

The following stimulants are prohibited, including both their optical (D- and L-) isomers where relevant:

Adrafinil, adrenaline,[4] amfepramone, amiphenazole, amphetamine, amphetaminil, benzphetamine, bromantan, carphedon, cathine,[5] clobenzorex, cocaine, cropropamide, crotetamide, cyclazodone, dimethylamphetamine, ephedrine,[6] etamivan, etilamphetamine, etilefrine, famprofazone, fenbutrazate, fencamfamin, fencamine, fenetylline, fenfluramine, fenproporex, furfenorex, heptaminol, isometheptene, levmethamfetamine, meclofenoxate, mefenorex, mephentermine, mesocarb, methamphetamine (D-), methylenedioxyamphetamine, methylenedioxymethamphetamine, *p*-methylamphetamine, methylephedrine,[6] methylphenidate, modafinil, nikethamide, norfenefrine, norfenfluramine, octopamine, ortetamine, oxilofrine, parahydroxyamphetamine, pemoline, pentetrazol, phendimetrazine, phenmetrazine, phenpromethamine, phentermine, prolintane, propylhexedrine, selegiline, sibutramine, strychnine, and other substances with a similar chemical structure or similar biological effect(s).[7]

[4]Adrenaline associated with local anesthetic agents or by local administration (e.g., nasal, ophthalmologic) is not prohibited.

[5]Cathene is prohibited when its concentration in urine is >5 μg/mL.

[6]Each of ephedrine and methylephedrine is prohibited when its concentration in urine is greater than 10 μg/mL.

[7]The following substances included in the 2006 Monitoring Program (bupropion, caffeine, phenylephrine, phenylpropanolamine, pipradol, pseudoephedrine, synephrine) are not considered as prohibited substances.

S7. NARCOTICS

The following narcotics are prohibited:

Buprenorphine, dextromoramide, diamorphine (heroin), fentanyl and its derivatives, hydromorphone, methadone, morphine, oxycodone, oxymorphone, pentazocine, and pethidine.

S8. CANNABINOIDS

Cannabinoids (e.g., hashish, marijuana) are prohibited.

S9. GLUCOCORTICOSTEROIDS

All glucocorticosteroids are prohibited when administered orally, rectally, intravenously, or intramuscularly. Their use requires a therapeutic use exemption approval.

Except as indicated below, other routes of administration require an abbreviated therapeutic use exemption.

Topical preparations when used for dermatological, aural/otic, nasal, buccal cavity and ophthalmologic disorders are not prohibited and do not require any form of therapeutic use exemption.

SUBSTANCES PROHIBITED IN PARTICULAR SPORTS

P1. ALCOHOL

Alcohol (ethanol) is prohibited *in-competition* only, in the following sports. Detection will be conducted by analysis of breath and/or blood. The doping violation threshold for each Federation is reported in parenthesis.

- Aeronautic (FAI) (0.20 g/L)
- Archery (FITA, IPC) (0.10 g/L)
- Automobile (FIA) (0.10 g/L)
- *Billiards (WCBS) (0.20 g/L)
- Boules (CMSB) (0.10 g/L)

IPC bowls

- Karate (WKF) (0.10 g/L)
- Modern pentathlon (UIPM) (0.10 g/L) for disciplines involving shooting
- Motorcycling (FIM) (0.10 g/L)
- Powerboating (UIM) (0.30 g/L)

P2. BETABLOCKERS

Unless otherwise specified, beta-blockers are prohibited *in-competition* only, in the following sports:

- Aeronautic (FAI)
- Archery (FITA, IPC) (also prohibited *out-of-competition*)
- Automobile (FIA)
- Billiards (WCBS)
- Bobsleigh (FIBT)
- Boules (CMSB, IPC bowls)
- Bridge (FMB)
- Chess (FIDE)
- Curling (WCF)
- Gymnastics (FIG)
- Motorcycling (FIM)
- Modern pentathlon (UIPM) for disciplines involving shooting
- Nine-pin bowling (FIQ)
- Sailing (ISAF) for match race helms only
- Shooting (ISSF, IPC) (also prohibited *out-of-competition*)

- Skiing/snowboarding (FIS) in ski jumping, freestyle aerials/halfpipe and snowboard halfpipe/big air
- Wrestling (FILA)

Beta-blockers include, but are not limited to, the following:

Acebutolol, alprenolol, atenolol, betaxolol, bisoprolol, bunolol, carteolol, carvedilol, celiprolol, esmolol, labetalol, levobunolol, metipranolol, metoprolol, nadolol, oxprenolol, pindolol, propranolol, sotalol, and timolol.

SPECIFIED SUBSTANCES[8]

"Specified substances"[8] are listed below:

- All inhaled beta-2 agonists, except clenbuterol
- Probenecid
- Cathine, cropropamide, crotetamide, ephedrine, etamivan, famprofazone, heptaminol, isometheptene, levmethamfetamine, and meclofenoxate
- *p*-Methylamphetamine, methylephedrine, nikethamide, norfenefrine,
  octopamine, ortetamine, oxilofrine, phenpromethamine, propylhexedrine, selegiline, and sibutramine
- Cannabinoids
- All glucocorticosteroids
- Alcohol
- All beta blockers

[8]"*The Prohibited List may identify specified substances which are particularly susceptible to unintentional anti-doping rule violations because of their general availability in medicinal products or which are less likely to be successfully abused as doping agents*." A doping violation involving such substances may result in a reduced sanction provided that the "... *Athlete can establish that the use of such a specified substance was not intended to enhance sport performance* ...."

FURTHER READING

Anderson, W. (2005). Human gene therapy: Scientific and ethical considerations. *J. Med. Philos.* **10,** 275–291.

Boone, K. (1988). Bad axioms in genetic engineering. *Hastings Center Report* **18,** August/September, 9–13.

Breivik, G. (1987). The doping dilemma: Some game theoretical and philosophical considerations. *Sportwissenshaft* **17**(1), 83–94.

Breivik, G. (1992). Doping games a game theoretical exploration of doping. *Int. Rev. Soc. Sport* **27,** 235–252.

Brown, M. (1980). Ethics, drugs, and sport. *J. Philos. Sport* **VII,** 15–23.

Brown, M. (1984a). Paternalism, drugs, and the nature of sports. *J. Philos. Sport* **XI,** 14–22.

Brown, M. (1984b). Comments on Simon and Fraleigh. *J. Philos. Sport* **XI,** 33–35.

Brown, M. (1990). "Practices and prudence." *J. Philos. Sport* **XVII,** 71–84.

Canadian Centre for Ethics in Sport. "Cornerstones of Community: Highlights of the National Survey of Nonprofit and Voluntary Organizations." Statistics Canada, Canada.

Canadian Centre for Ethics in Sport web site www.cces.ca

Chapman, A. R, and Frankel, M. S. (2003). "Designing Our Descendants: The Promises and Perils of Genetic Modifications." The Johns Hopkins University Press, Baltimore.

Feezell, R. (1986). Sportsmanship. *J. Philos. Sport* **XIII,** 1–13.

Feezell, R. (1988). On the wrongness of cheating and why cheaters can't play the game. *J. Philos. Sport* **XV,** 57–68.

Feinberg, J. (1977). Legal paternalism. *Can. J. Philos.* **1,** 106–124.

Fost, N. (1986). Banning drugs in sport: A skeptical view. *Hasting Centre Report,* **August,** 5–10.

Friedmann, T. (1983). Gene Therapy: Fact and Fiction. *In* "Biology's New Approaches to Disease," pp. 91–96. Cold Spring Harbor Laboratory, Cold Spring Harbor.

Friedmann, T. (1997). Overcoming the obstacles to gene therapy. *Sci. Am.* **276,** 95–101.

Friedmann, T. (1999). "The Development of Human Gene Therapy." Cold Spring Harbor Press, Cold Spring Harbor.

Friedmann, T., and Roblin, R. (1972). Gene therapy for human genetic disease? *Science* **175,** 949–955.

Friedmann, T., and Koss, J. O. (2001). Gene transfer and athletics: An impending problem. *Mol. Ther.* **3,** 819–820.

Gardner, R. (1989). On performance-enhancing substances and the unfair advantage argument. *J. Philos. Sport* **XVI,** 59–73.

Gardner, R. (1990). "Performance-Enhancing Substances in Sport: An Ethical Study." Ph.D. thesis. University Microfilms International, Purdue University, Michigan.

Gaylin, W. (1984). Feeling good and doing better. *In* "Feeling Good and Doing Better: Ethics and Nontherapeutic Drug Use" (T. H. Murray, W. Gaylin, and R. Macklin, eds.), pp. 1–10. Humana Press, New Jersey.

Goldman, B. (1987). "Death in The Locker Room." The Body Press, Tucson.

Hoberman, J. (1988). Sport and the technological image of man. *In* "Philosophic Inquiry in Sport" (W. J. Morgan and K. V. Meier, eds.). Human Kinetics Publishers Inc., Champaign.

Hoberman, J. (1992). Mortal Engines.

Hughes, R., and Coakley, J. (1991). Positive deviance among athletes: The implications of overconformity to the sport ethic. *Sociol. Sport J.* **8,** 307–325.

HUGO Ethics Committee (2001). Statement on gene therapy research. *Eubios J. Asian Int. Bioeth.* **11,** 98–99.

Hyland, D. (1988). Competition and friendship. *In* "Philosophic Inquiry in Sport" (W. J. Morgan and K. V. Meier, eds.). Human Kinetics Publishers Inc., Champaign.

Hyland, D. (1990). "Philosophy of Sport." Paragon House, New York.

Juengst, E., and Parens, E. (2003). Germ-line dancing: Definitional considerations for policy makers. *In* "Designing Our Descendants: The Promises and Perils of Genetic Modifications." Johns Hopkins University Press, Baltimore.

Juengst, E. T. (2004). Enhancement uses of medical technology. "Encyclopedia of Bioethics," 3rd Ed., Vol. 2. Macmillan Reference, New York, USA.

Juengst, E. T. (2004). Anti-aging research and the limits of medicine. *In.* "Fountain of Youth: Cultural, Scientific, and Ethical Perspectives on a Biomedical Goal." Oxford University Press, New York.

Kretchmar, R. (1972). Ontological possibilities: Sport as play. *Philos. Exchange* **1,** 113, 124.

Kretchmar, R. (1983). Ethics and sport: An overview. *J. Philos. Sport* **X,** 21–32.

Kretchmar, R. (1988). From test to contest: An analysis of two kinds of counterpoint in sport. *In* "Philosophic Inquiry in Sport" (W. Morgan and K. Meier, eds.), pp. 223–230. Human Kinetics Publishers Champaign.

Lavin, M. (1987). Sports and drugs: Are the current bans justified? *J. Philos. Sport* **XIV,** 35–43.

Mac Intyre, A. (1984). "After Virtue," 2nd Ed. University of Notre Dame Press, Notre Dame.

Mehlman, M. J. (2003). "Wondergenes: Genetic Enhancement and the Future of Society." Indiana University Press, Bloomington.

Miah, A. (2004). "Genetically Modified Athletes: Biomedical Ethics, Gene Doping, and Sport." Routledge, London.

Mill, J. (1978). "On Liberty." Hackett Publishers, Indianapolis.

Murray, T. H. (2004). Bioethics in sports. "Encyclopedia of Bioethics," 3rd Ed., Vol. 4. Macmillan Reference, New York.

Murray, T. H. (1983). "The Coercive Power of Drugs in sport." *The Hasings Center Report* **13,** 24–30.

Murray, T. H. (1984). Drugs, sports, and ethics. *In* "Feeling Good and Doing Better: Ethics and Nontherapeutic Drug Use" (T. H. Murray, W. Gaylin, and R. Macklin, eds.), pp. 107–126. Humana Press, New Jersey.

Murray, T. H. (1986a). Human growth hormone in sports: No. Minneapolis. *Physician Sportsmed.* **14**(5), May, 29.

Murray, T. H. (1986b). Drug testing and moral responsibility. Minneapolis. *Physician Sportsmed.* **14**(1), November, 47–48.

Murray, T. (1987). The ethics of drugs in sport. *In* "Drugs and Performance in Sports" (R. H. Strausss, ed.), pp. 11–21. W. B. Sunders, Philadelphia.

Palmer, C. (2000). Spin doctors and sportsbrokers. *Int. Rev. Sociol. Sport (London)* **35**(3), 364–377.

Parens, E., Chapman, A.R, and Press, N. (2005). "Wrestling with Behavioral Genetics: Science, Ethics, and Public Conversation." Johns Hopkins University Press, Baltimore.

Parfit, D. (1984). "Reasons and Persons." Oxford University Press, Oxford.

Schneider, Angela (2006). Performance Enhancement in sport: An athlete's perspective. *In* "Performance-Enhancing Technologies in Sports: Ethical, Conceptual, and Scientific Issues" (Thomas Murray *et al.*, eds.). John Hopkins University Press, Baltimore. (Forthcoming).

Schneider, A. J., and Butcher, R. B. (1993). "The Ethical Rationale for Drug-Free Sport," p. 155. Canadian Centre for Drug-Free Sport (pub.), Ottawa, Ontario, Canada.

Schneider, A. J., and Butcher, R. B. (1994). "Why Olympic Athletes Should Avoid the Use and Seek the Elimination of Performance-Enhancing Substances and Practices in the Olympic Games." *J. Philos. Sport* **XX, XXI,** pp. 64–81. Human Kinetics (pub.), ISSN 0094–8705.

Schneider, A. J., and Butcher, R. B. (2000a). A philosophical overview of the arguments on banning doping in sport, Chapter 13 in Part V: The scientific manufacture of winners. *In* "Values in Sport: Elitism, Nationalism, Gender Equality and the Scientific Manufacture of Winners" (T. Tannsjo and C. Tamburrini, eds.), pp. 185–200. E & FN Spon, London and New York.

Schneider, A. J., and Butcher, R. B. (2000b). An ethical analysis of drug testing. *In* "Doping in Elite Sport: The Politics of Drugs in the Olympic Movement" (W. Wilson and E. A. Derse, eds.), Chapter 5, pp. 102–124. Human Kinetics, Champaign, IL. ISBN 0–7360–0329–0.

Schneider, A. J., and Butcher, R. B. (2003). Community sport, community choice: The ethical challenges of community sport. *In* "The Sport We Want: Essays on Current Issues in Community Sport in Canada," pp. 45–58. The Canadian Centre for Ethics in Sport (pub.), Ottawa, Canada.

Simon, R. L. (1984a). Good competition and drug-enhanced performance. *J. Philos. Sport* **XI,** 6–13.

Simon, R. L. (1984b). Response to Brown and Fraleigh. *J. Philos. Sport* **XI,** 30–32.

Simon, R. L. (1991). "Fair Play: Sports, Values and Society. Boulder." Westview Press, CO.

Voet, W. (2001). *In* "Breaking the Chain: How Drugs Destroyed a Sport," p. 128. Yellow Jersey Press, ISBN 0224060562.

Sweeney, L. (2004). Gene doping. *Sci. Am.*, **July.**

Wolfe, M. L. (1989). Correlates of adaptive and maladaptive musical performance anxiety. *Med. Probl. Performing Artists* **March,** 49–56.

World Anti-Doping Agency. World Anti–Doping Code, www.wada.org.

World Anti-Doping Agency. Athlete's Guide, www.wada.org.

Index

H